Victor Manuel Menendez Flores

**Developing Nanoparticles for decomposition of toxic compounds**

Victor Manuel Menendez Flores

# Developing Nanoparticles for decomposition of toxic compounds

Synthesis of Semiconductor Nanoparticles Applied in Photocatalysis for the Degradation of Pollutants in Aqueous and Gas Phase

Südwestdeutscher Verlag für Hochschulschriften

**Impressum/Imprint (nur für Deutschland/ only for Germany)**
Bibliografische Information der Deutschen Nationalbibliothek: Die Deutsche Nationalbibliothek verzeichnet diese Publikation in der Deutschen Nationalbibliografie; detaillierte bibliografische Daten sind im Internet über http://dnb.d-nb.de abrufbar.

Alle in diesem Buch genannten Marken und Produktnamen unterliegen warenzeichen-, marken- oder patentrechtlichem Schutz bzw. sind Warenzeichen oder eingetragene Warenzeichen der jeweiligen Inhaber. Die Wiedergabe von Marken, Produktnamen, Gebrauchsnamen, Handelsnamen, Warenbezeichnungen u.s.w. in diesem Werk berechtigt auch ohne besondere Kennzeichnung nicht zu der Annahme, dass solche Namen im Sinne der Warenzeichen- und Markenschutzgesetzgebung als frei zu betrachten wären und daher von jedermann benutzt werden dürften.

Verlag: Südwestdeutscher Verlag für Hochschulschriften GmbH & Co. KG
Dudweiler Landstr. 99, 66123 Saarbrücken, Deutschland
Telefon +49 681 37 20 271-1, Telefax +49 681 37 20 271-0
Email: info@svh-verlag.de
Zugl.: Hannover, Leibniz Universität Hannover, Diss., 2010

Herstellung in Deutschland:
Schaltungsdienst Lange o.H.G., Berlin
Books on Demand GmbH, Norderstedt
Reha GmbH, Saarbrücken
Amazon Distribution GmbH, Leipzig
**ISBN: 978-3-8381-1953-3**


**Imprint (only for USA, GB)**
Bibliographic information published by the Deutsche Nationalbibliothek: The Deutsche Nationalbibliothek lists this publication in the Deutsche Nationalbibliografie; detailed bibliographic data are available in the Internet at http://dnb.d-nb.de.

Any brand names and product names mentioned in this book are subject to trademark, brand or patent protection and are trademarks or registered trademarks of their respective holders. The use of brand names, product names, common names, trade names, product descriptions etc. even without a particular marking in this works is in no way to be construed to mean that such names may be regarded as unrestricted in respect of trademark and brand protection legislation and could thus be used by anyone.

Publisher: Südwestdeutscher Verlag für Hochschulschriften GmbH & Co. KG
Dudweiler Landstr. 99, 66123 Saarbrücken, Germany
Phone +49 681 37 20 271-1, Fax +49 681 37 20 271-0
Email: info@svh-verlag.de

Printed in the U.S.A.
Printed in the U.K. by (see last page)
**ISBN: 978-3-8381-1953-3**


Copyright © 2010 by the author and Südwestdeutscher Verlag für Hochschulschriften GmbH & Co. KG and licensors
All rights reserved. Saarbrücken 2010

# Index

Index..................................................................................................................i
Index of figures ............................................................................................v
Index of tables ............................................................................................xi
Acknowledgements.....................................................................................xii
Symbols and abbreviations.........................................................................xiii
Kurzfassung................................................................................................xvi
Abstract .....................................................................................................xvii

**I. Introduction and Problem Statement** ................................................. 1

**II. Theory**................................................................................................ 4

II.1. Fundamentals..................................................................................... 4
II.1.1. Semiconductor-interface behavior in absence of redox systems........ 6
II.1.2. Semiconductor-interface behavior in presence of redox systems ..... 9
II.1.3. Behavior of illuminated semiconductor-interface............................. 13
II.1.4. Photocatalytic reactions by charge transfer at semiconductor nanoparticles .... 19
II.1.5. Quantum size effect........................................................................ 22
II.2. Heterogeneous photocatalysis.......................................................... 23
II.2.1. Mechanisms................................................................................... 24
II.2.2. Stability problems .......................................................................... 26
II.3. New materials for photocatalysis ..................................................... 28
II.3.1. Photodeposited photocatalysts ...................................................... 28
II.3.2. Doped material .............................................................................. 31
II.3.3. Heat treatment for structure modifications ..................................... 32
II.4. Diverse photocatalytic test systems ................................................. 33
II.4.1. Self cleaning effect ........................................................................ 34
II.4.2. Photocatalytic decomposition of DCA on bare $TiO_2$ in aqueous phase ............. 36
II.4.3. Photocatalytic gas phase decomposition ....................................... 38

# Index

## III. Materials and Methods .................................................................................. 41

III.1. For deposition synthesis of Ag on $TiO_2$ ................................................................ 41
III.2. For the S-doped $TiO_2$-$Fe^{+3}$ photocatalyst ............................................................. 41
III.3. For the synthesis of indium selenide ................................................................... 42
III.4. For the synthesis of beta gallium oxide ............................................................... 42
III.5. Set-up for photoactivity tests in aqueous phase, DCA degradation ..................... 43
III.6. Measurements of pH, chloride ions and total organic carbon (TOC) ................... 45
III.7. Set-up for photoactivity tests in gas phase .......................................................... 45
III.8. Set-up for carrying out photocurrent measurements ........................................... 47
III.9. Set-up for cyclic voltammetry and Mott-Schottky measurements ....................... 47

## IV. Results and Discussion .................................................................................. 49

IV.1. Photonic efficiency calculation ............................................................................ 49
IV.1.1. Photonic efficiency calculation for UV-A light .................................................. 49
IV.1.2. Photonic efficiency calculation modified for visible light intensity .................... 49
IV.1.3. Photonic efficiency calculation for the gas phase system ................................ 51
IV.2. Degussa P25 as a standard photocatalyst .......................................................... 53
IV.2.1. Degradation of DCA with bare Degussa P25 and its effect after washing ........ 53
IV.2.2. Degradation of $NO_x$ with Degussa P25 ........................................................... 55
IV.2.3. Degradation of acetaldehyde with Degussa P25 .............................................. 57
IV.3. Durability of Ag-$TiO_2$ Photocatalysts Assessed for the Degradation of Dichloroacetic Acid .................................................................................................... 60
IV.3.1. Preparation of Ag-$TiO_2$ and colloidal $TiO_2$ photocatalyst ................................ 60
IV.3.2. Analysis and characterization of the Ag-$TiO_2$ photocatalysts ......................... 61
IV.3.3. Degradation of DCA with photodeposited silver on Degussa P25 ................... 65
IV.3.4. Degradation of DCA with self prepared colloidal $TiO_2$ particles ...................... 67
IV.3.5. Total organic carbon Ag-$TiO_2$ results ............................................................... 67
IV.3.6. Photonic efficiency Ag-$TiO_2$ results .................................................................. 71
IV.3.7. Conclusions ...................................................................................................... 75

# Index

IV.4. Photocatalytic activities under visible light by S-doped $TiO_2Fe^{3+}$ photocatalyst .. 76
IV.4.1. Preparation of S-doped $TiO_2$-$Fe^{+3}$ photocatalyst............................................. 76
IV.4.2. Analysis and characterization of the S-doped $TiO_2$-$Fe^{3+}$ photocatalyst............ 76
IV.4.3. Photocatalytic decomposition of DCA on S-doped $TiO_2$-$Fe^{3+}$ material............ 82
IV.4.4. Degradation of DCA with S-doped $TiO_2$-$Fe^{3+}$ at pH 3 under visible light.......... 84
IV.4.5. Different pH values for degradation of DCA with S-doped $TiO_2$-$Fe^{3+}$ ............... 85
IV.4.6. Stability test by a consecutively DCA degradation reactions with S-doped $TiO_2$-$Fe^{3+}$ at diverse pH conditions.................................................................................. 86
IV.4.7. Comparison of commercial photocatalysts under UV-A and visible light.......... 89
IV.4.8. Photocatalytic decomposition of $NO_x$ or acetaldehyde on S-$TiO_2$-$Fe^{3+}$ nanoparticles under visible light in a gas phase reactor............................................. 91
IV.4.9. Photonic efficiency S-$TiO_2$-$Fe^{3+}$ results................................................................ 94
IV.4.10. Conclusions..................................................................................................... 96
IV.5. Solid state synthesis and characterization of $In_2Se_3$ nanoparticles deposited by heat treatment as a film electrode................................................................................. 98
IV.5.1. Synthesis development of $In_2Se_3$ and $In_6Se_7$ nanocrystals ............................ 100
IV.5.2. Analysis and characterization of the synthesized indium selenide material ... 101
IV.5.3. Mott-Schottky study of the $In_2Se_3$ electrode..................................................... 111
IV.5.4. Current and photocurrent measurements of $In_2Se_3$ electrode........................ 114
IV.5.5. Conclusions................................................................................................... 122
IV.6. Solid state synthesis of $\beta$-$Ga_2O_3$ by heat treatment and characterized as a film electrode or powder showing photocatalytic improvement decomposing acetaldehyde ...................................................................................................................... 123
IV.6.1. Synthesis of gallium acetate........................................................................ 124
IV.6.2. Synthesis and characterization of $\beta$-$Ga_2O_3$ ................................................. 124
IV.6.3. Photocatalytic decomposition of acetaldehyde on $\beta$-$Ga_2O_3$ nanoparticles under UV-A light ................................................................................................................ 133
IV.6.4. Preparation of $Ga_2O_3$ electrode .................................................................. 134
IV.6.5. Characterization of the $Ga_2O_3$ electrode..................................................... 134
IV.6.6. Mott-Schottky study of the $Ga_2O_3$ electrode ............................................... 136
IV.6.7. Current and photocurrent measurements of $Ga_2O_3$ electrode....................... 137

IV.6.8. Conclusions .................................................................................. 141

**V. General conclusions** ..................................................................... 142

**VI. Summary** ....................................................................................... 143

**VII. References** .................................................................................. 148

**VIII. Annex 1** ...................................................................................... 158
VIII.1. Halogen lamp spectrum for visible light emission ........................... 158

# Index

## Index of figures

Figure II-1. Position of energy bands at the surface of various semiconductors in aqueous electrolytes at pH 0 [6] (modified). ............ 4

Figure II-2. Principle mechanism of photocatalysis [24]. ............ 6

Figure II-3. Potential distribution at the semiconductor-electrolyte interface [6]. ............ 8

Figure II-4. Electron energies of a redox system vs. density of states: a) $E^0_{red}$ for occupied states, $E^0_{ox}$ for empty states, A as electron affinity and I as ionization energy of the redox system; b) the corresponding distribution functions at $C_{ox}=C_{red}$; c) the distribution functions at $C_{ox}<<C_{red}$ [6]. ............ 11

Figure II-5. Electron energies of a semiconductor electrode in contact with a redox system [6] ............ 13

Figure II-6. Charge transfer at the semiconductor-solution interface under illumination [6]. ............ 14

Figure II-7. Charge carrier transfer at large (A) and small (B) semiconductor particles in the presence of an electron donor D and an acceptor A [26]. ............ 20

Figure II-8. Mechanism of reaction on the surface of $TiO_2$ with photodeposited Ag [35]. ............ 25

Figure II-9. Mechanism of the reaction on the rutile S-doped $TiO_2$-$Fe^{3+}$ photocatalyst under UV (A) and visible (B) light irradiation [37]. ............ 26

Figure II-10. Photocatalytic applications [62]. ............ 34

Figure II-11. Before irradiation (A) and after irradiation (B) [62]. ............ 34

Figure II-12. Superhydrophilicity occurs under light irradiation [62]. ............ 35

Figure II-13. Hydrophobic (A) and after light irradiation hydrophilic and oleophilic (B) (Photo, Königs). ............ 36

Figure II-14. Mechanism of DCA degradation of the photocatalytic reactor set up. ............ 37

Figure II-15. Schematic diagram of the DCA degradation. ............ 38

Figure II-16. $NO_x$ and $SO_x$ can be removed from the environment through photocatalysis [62]. ............ 39

Figure III-1. Photocatalytic reactors (A) 50 mL and (B) 120 mL used for DCA degradation. Latter one designed to introduce a selective chloride electrode to follow $Cl^-$ in-situ. ............ 43

Figure III-2. Schematic of the photocatalytic reactor set-up. ............ 44

Figure III-3. Picture of the photocatalytic pH-stat system. ............ 44

Figure III-4. Schematic presentation of the photocatalytic reactor set-up for gas phase. ............ 46

Figure III-5. Picture of the photocatalytic $NO_x$ (A) and acetaldehyde (B) gas phase set-up systems. ............ 46

Figure III-6. Schematic presentation of the photocurrent set-up system. ............ 47

Figure III-7. Schematic presentation of the set-up system for voltammetry measurements and Mott-Schottky diagrams construction. ............ 48

Figure IV-1. Relation between the S-doped $TiO_2$-$Fe^{3+}$ photocatalyst (---) and a xenon lamp (—)spectra to obtain the common area (53.545 a.u.; ▦) of the available visible light photons from 420 nm (cut off filter(···)) to the absorbing material limit 555 nm. For the UV light experiments the cut off filter 320 nm (—) was used. ............ 51

Figure IV-2. The shared area between the S-doped $TiO_2$-$Fe^{3+}$ diffuse reflectance spectrum (—) and the visible light source spectra with or without any filter from 400 nm to 555 nm (▦) (29.273) was obtain as a factor for further calculations of available visible light photons. The intensity without any filter (■) or with a glass filter (●) was 0.793 mW/cm²; under a polycarbonate filter (▲) 0.642 mW/cm² and under a green filter (▼) 0.509 mW/cm². ............ 52

# Index

Figure IV-3. Degradation of DCA (shown as release of H⁺) using the photocatalyst Degussa P25 at pH 3 in 3 consecutive runs, 1st run (—), 2nd run (---), 3rd run (⋯) and a 3rd run (-⋅-) with addition of 4 mM Cl⁻ before the run started. The slope (–). used for the determination of the photonic efficiency of each run with $I \approx 3.39 \times 10^{-2}$ Einstein $L^{-1}h^{-1}$. The photocatalyst loading was 0.5 g/L. ............................................................................................................................................ 54

Figure IV-4. Degradation of DCA (shown as release of H⁺) using the photocatalyst Degussa P25 at pH 3 in 3 consecutive runs with intermittent washing between the runs, 1st run (—), 2nd run (---), 3rd run (⋯) and slope (–) used for the determination of the photonic efficiency of each run with $I \approx 3.39 \times 10^{-2}$ Einstein $L^{-1}h^{-1}$. The photocatalyst loading was 0.5 g/L. .......................................... 54

Figure IV-5. Comparison of 1 ppm NOx decomposition with 4 g P25 pressed powder photocatalyst applying 1 mW/cm² UV-A light without any filter. The reaction was followed by measuring NO$_x$ (●); NO (■); and NO$_2$ (▲). ................................................................................................. 55

Figure IV-6. Comparison of 1ppm NOx decomposition with 4 g P25 pressed powder photocatalyst applying visible light photons under a Pilkington green filter, polycarbonate and without any filter. The reaction was followed by measuring NO$_x$ (●); NO (■); and NO$_2$ (▲). ............................... 57

Figure IV-7. Comparison of 1ppm Acetaldehyde (—) degradation with 4 g P25 pressed powder photocatalyst applying 1 mW/cm² UV light photons without any filter. ...................... 58

Figure IV-8. Comparison of 1 ppm Acetaldehyde (—) degradation with 4 g P25 pressed powder photocatalyst applying visible light photons under a Pilkington green filter, polycarbonate and without any filter. ........................................................................................................ 59

Figure IV-9. EDXS-spectrum of Degussa P25 prior to the photodeposition of silver or the photodegradation of DCA. .................................................................................................. 61

Figure IV-10. EDXS spectrum of 0.35 Ag-TiO$_2$ photocatalyst prior to the photodegradation of DCA. 62

Figure IV-11. EDXS spectrum of 0.35 Ag-TiO$_2$ photocatalyst after photodegradation of DCA in three consecutive runs. .............................................................................................................. 62

Figure IV-12. Electron microscopy analysis of 0.35 Ag-TiO$_2$ photocatalyst particles before the DCA photodegradation reaction (A) STEM image of particles, (B) corresponding Ti x-ray map and (C) corresponding Ag x-ray map. ............................................................................................. 63

Figure IV-13. Electron microscopy analysis of 0.35 Ag-TiO$_2$ photocatalyst particles after the DCA photodegradation reaction (A) STEM image of particles, (B) corresponding Ag x-ray map (C) corresponding Cl x-ray map. ............................................................................................. 63

Figure IV-14. TEM images of 0.35 Ag-TiO$_2$ photocatalyst (A) before and (B) after recycle. .............. 65

Figure IV-15. Degradation of DCA (shown as release of H⁺) using the photocatalyst 0.35 Ag-TiO$_2$ at pH 3 in 3 consecutive runs, 1st run (—), 2nd run (---), 3rd run (⋯), a 3rd run (-⋅-) with addition of 4 mM Cl⁻ before the run started and slope (–) used for the determination of the photonic efficiency of each run with $I \approx 3.39 \times 10^{-2}$ Einstein $L^{-1}h^{-1}$. The photocatalyst loading was 0.5 g/L. ............. 65

Figure IV-16. Degradation of DCA (shown as release of H⁺) using the photocatalyst 0.35 Ag-TiO$_2$ at pH 3 in 3 consecutive runs with intermittent washing between runs, 1st run (—), 2nd run (---), 3rd run (⋯) and slope (–) used for the determination of the photonic efficiency with $I \approx 3.39 \times 10^{-2}$ Einstein $L^{-1}h^{-1}$. The photocatalyst loading was 0.5 g/L. ........................................................... 66

Figure IV-17. Degradation of DCA (shown as release of H⁺) using the prepared colloidal TiO$_2$ photocatalyst at pH 3 in 3 consecutive runs with intermittent washing between runs, 1st run (—), 2nd run (---), 3rd run (⋯), a 3rd run (-⋅-) with addition of 4 mM Cl⁻ before the run started and slope

(—) used for the determination of the photonic efficiency of each run with $I \approx 3.39 \times 10^{-2}$ Einstein $L^{-1}h^{-1}$. The photocatalyst loading was 0.5 g/L. ............................................................... 67

Figure IV-18. Removal of DCA (TOC removal) after 4 hours of illumination using the photocatalyst P25-TiO$_2$ at pH 3 in 3 consecutive runs (1$^{st}$ run ■, 2$^{nd}$ run ■ and 3$^{rd}$ run □), the H$^+$ production efficiency after 4 hours of illumination is also shown (1$^{st}$ run ▨, 2$^{nd}$ run ▨ and 3$^{rd}$ run ▨) in set A, with chloride ion addition before the third run (set B), and with intermittent washing between runs (set C). The photocatalyst loading was 0.5 g/L and the light intensity $I \approx 3.39 \times 10^{-2}$ Einstein $L^{-1}h^{-1}$. ............................................................................................................................. 68

Figure IV-19. Removal of DCA (TOC removal) after 4 hours of illumination using the photocatalyst 0.35 Ag-TiO$_2$ at pH 3 in 3 consecutive runs (1$^{st}$ run ■, 2$^{nd}$ run ■ and 3$^{rd}$ run □), the H$^+$ production efficiency after 4 hours of illumination is also shown (1$^{st}$ run ▨, 2$^{nd}$ run ▨ and 3$^{rd}$ run ▨) in set A, with chloride ion addition in set B before the third run and with intermittent washing between runs (set C). The photocatalyst loading was 0.5 g/L and the light intensity $I \approx 3.39 \times 10^{-2}$ Einstein $L^{-1}h^{-1}$. ............................................................................................................................. 69

Figure IV-20. Removal DCA (TOC removal) after 4 hours of illumination using the colloidal-TiO$_2$ photocatalyst at pH 3 in 3 consecutive runs (1$^{st}$ run ■, 2$^{nd}$ run ■ and 3$^{rd}$ run □) the H$^+$ production efficiency after 4 hours of illumination is also shown (1$^{st}$ run ▨, 2$^{nd}$ run ▨ and 3$^{rd}$ run ▨) in set A and with chloride ion addition before the third run (set B). The photocatalyst loading was 0.5 g/L and the light intensity $I \approx 3.39 \times 10^{-2}$ Einstein $L^{-1}h^{-1}$. ............................................. 70

Figure IV-21. Effect of recycling Ag-TiO$_2$ photocatalysts on observed photonic efficiencies at pH 3 and pH 10 (1$^{st}$ run ■, 2$^{nd}$ run ■ and 3$^{rd}$ run □), degradation of 1mM DCA without any extra addition of chloride ions or washing technique performed. The photocatalyst concentration was 0.5 g/L with different silver loadings (atom%). The light intensity at pH 3 was $I \approx 3.39 \times 10^{-2}$ Einstein $L^{-1}h^{-1}$ and at pH 10 was $I \approx 3.74 \times 10^{-2}$ Einstein $L^{-1}h^{-1}$. .................................. 72

Figure IV-22. Effect of recycling Ag-TiO$_2$ photocatalysts on observed TOC removal at pH 3 and pH 10 (1$^{st}$ run ■, 2$^{nd}$ run ■ and 3$^{rd}$ run □), degradation of 1 mM DCA without any extra addition of chloride ions or washing technique performed. The light intensity at pH 3 was $I \approx 3.39 \times 10^{-2}$ Einstein $L^{-1}h^{-1}$ and at pH 10 was $I \approx 3.74 \times 10^{-2}$ Einstein $L^{-1}h^{-1}$. ....................................... 74

Figure IV-23. XRD of PC500 from Millenium pure anatase (A) and S-doped TiO$_2$-F$^{3+}$ (B) material... 78

Figure IV-24. Reflectance function (A) and normalized Reflectance function (B) of P25 (—), anatase PC500 (—) and S-doped TiO$_2$-Fe$^{3+}$ (---). ......................................................................... 79

Figure IV-25. The band gap of P25 (—) and anatase PC500 (—) is 3.22 eV but for S-doped TiO$_2$-Fe$^{3+}$ is 3.35 eV (---). ............................................................................................................. 79

Figure IV-26. Zero point of charge comparison measurements of (5.0g/L) P25 (—■—) and S-TiO$_2$-Fe$^{3+}$ (—■—) but also particle size (—●—) of S-TiO$_2$-Fe$^{3+}$ at different pH values in aqueous phase suspensions. ............................................................................................................. 81

Figure IV-27. TEM image of S-doped TiO$_2$-Fe$^{3+}$ photocatalyst after preparation. ............................ 81

Figure IV-28. S-doped TiO$_2$-Fe$^{3+}$ photocatalyst. ................................................................................ 82

Figure IV-29. Decomposition of 1.7 mMol DCA under 0.896 mW/cm$^2$ visible light (420 nm cut-filter) at pH3 with S-doped TiO$_2$-Fe$^{3+}$ photocatalyst using 0.5 g/L and 5.0 g/L followed by the release or measurement of H$^+$(---); ............................................................................................................. 83

Figure IV-30. Decomposition of 1.7 mMol DCA under 30 mW/cm$^2$ UV-light (320 nm cut-filter) at pH3 with S-doped TiO$_2$-Fe$^{3+}$ photocatalyst using 0.5 g/L and 5.0 g/L followed by the release or measurement of H$^+$(---); Cl$^-$(--▲--); TOC (--■--) and H$^+$(—); Cl$^-$(—▲—); TOC (—■—) respectively. ................................................................................................................................. 84

Figure IV-31. Photocatalytic activity by the degradation of 1 mMol DCA followed by the released .... 85

Figure IV-32. Photocatalytic activity degradation of 1 mMol DCA with 5.0 g/L S-doped $TiO_2$-$Fe^{3+}$ photocatalyst under 0.896 mW/$cm^2$ visible light (420 nm filter) by the $H^+$ release and compared at different pH conditions, pH 3 (—); pH 7 (---) and pH 9 (•••)................................................ 86

Figure IV-33. A consecutively repeated set of runs 1st run (—), 2nd run (---) and 3rd run (•••) was performed as a stability test for 5.0g/L S-doped $TiO_2$-$Fe^{3+}$ photocatalyst under 0.896 mW/$cm^2$ visible light (420 nm filter) at pH 3, during 16 hours each 1 mMol DCA degradation reaction, followed by the released of $H^+$................................................................................................ 87

Figure IV-34. A consecutively repeated set of runs 1st run (—), 2nd run (---) and 3rd run (•••) was performed as a stability test for 5.0g/L S-doped $TiO_2$-$Fe^{3+}$ photocatalyst under 0.896 mW/$cm^2$ visible light (420 nm filter) at pH 7, during 16 hours each 1 mMol DCA degradation reaction, followed by the released of $H^+$ at pH 7. ................................................................................. 87

Figure IV-35. A consecutively repeated set of runs 1st run (—), 2nd run (---) and 3rd run (•••) was performed as a stability test for 5.0g/L S-doped $TiO_2$-$Fe^{3+}$ photocatalyst under 0.896 mW/$cm^2$ visible light (420 nm filter) at pH 9, during 16 hours each 1mMol DCA degradation reaction, followed by the released of $H^+$................................................................................................. 88

Figure IV-36. Comparison of 5.0g/L photocatalysts EDXS spectra of bare anatase and S-doped $TiO_2$-$Fe^{3+}$ before any reaction. And after stability test performed by the consecutively 1mMol DCA degradation runs with S-doped $TiO_2$-$Fe^{3+}$, 1st run , 2nd run and 3rd run under under 0.896 mW/$cm^2$ visible light, at pH 3. Figures inserted correspond to the complete spectrum of each analysis. ............................................................................................................................................ 89

Figure IV-37. Photocatalytic activity comparison by the degradation of 1 mMol DCA between 5.0g/L of P25 (—), S-doped $TiO_2$-$Fe^{3+}$ (---) and Kronos (•••) photocatalysts under 30 mW/$cm^2$ UV-light (320 nm filter) resulting the S-doped $TiO_2$-$Fe^{3+}$ (⋯) as the only photoactive photocatalyst under 0.896 mW/$cm^2$ visible-light.(420 nm filter) at pH 3. ........................................................ 90

Figure IV-38. Comparison of 1 ppm NOx decomposition with 4 g S-doped $TiO_2$-$Fe^{3+}$ pressed powder photocatalyst applying visible light photons under a Pilkington green filter, polycarbonate and without any filter. The reaction was followed by measuring $NO_x$ (●); NO (■); and $NO_2$ (▲). .... 92

Figure IV-39. Comparison of 0.92 ppm Acetaldehyde (—) degradation with 4 g S-doped $TiO_2$-$Fe^{3+}$ pressed powder photocatalyst applying visible light photons under a Pilkington green filter, polycarbonate and without any filter. ................................................................................................ 94

Figure IV-40. Comparison between the reaction rates decomposition of the gas phase systems applying the photocatalyst P25 for decomposing ($NO_x$□; NO■; acetaldehyde■) and the photocatalyst S-$TiO_2$ ($NO_x$▨; NO▨; acetaldehyde▨) under different intensities of visible light. ................................................................................................................................................... 95

Figure IV-41. Comparison of 1ppm Acetaldehyde (—) degradation with 4 g P25 pressed powder photocatalyst applying visible light photons under a Pilkington green filter, polycarbonate and without any filter. ........................................................................................................................... 96

Figure IV-42. XRD pattern of the synthesized $In_2Se_3$ material (* peaks correspond to $In_2Se_3$ after WinXPOW database)................................................................................................................... 102

Figure IV-43. $In_2Se_3$ SEM images. The structure in (A) (—) represents 30 μm, in (B) (—) 20 μm and in (C) (—) 10 μm. ........................................................................................................................ 102

Figure IV-44. Surface of a $In_2Se_3$ slice................................................................................................. 103

Figure IV-45. EDX spectrum of $In_2Se_3$. ............................................................................................... 103

Figure IV-46. XRD pattern of the synthesized $In_6Se_7$ material (* peaks correspond to $In_6Se_7$ after WinXPOW database)................................................................................................................... 104

viii

# Index

Figure IV-47. In$_6$Se$_7$ SEM image. ............................................................................................... 105
Figure IV-48. EDX spectrum of In$_6$Se$_7$. .................................................................................... 105
Figure IV-49. XRD of the synthesized In$_2$Se$_3$ powder after one hour of calcination at 500°C (* peaks correspond to In$_2$Se$_3$ and ° peaks to In$_2$O$_3$ after WinXPOW database). ................... 106
Figure IV-50. Reflectance function (A) and normalized Reflectance function (B) of anatase PC500 (—), rutile R15 (---) and In$_2$Se$_3$(•••). ........................................................................ 107
Figure IV-51. (A) In$_2$Se$_3$ indirect band gap determination and (B) In$_2$Se$_3$ direct band gap determination. ................................................................................................................... 108
Figure IV-52. In$_2$Se$_3$ powder scan microscopy. ...................................................................... 109
Figure IV-53. Elemental electron microscopy analysis of In$_2$Se$_3$ particles(A), corresponding indium mapping (B) and corresponding selenium mapping (C). .................................................... 110
Figure IV-54. In$_2$Se$_3$ EDXS after 500°C calcination. ............................................................. 111
Figure IV-55. Mott-Schottky diagram from the 0.25 cm$^2$ In$_2$Se$_3$ electrode. Experiment performed in aqueous 0.1 M KCl solution at pH 7. With frequency range modulated between 100 Hz and 1 kHz. ................................................................................................................................ 113
Figure IV-56. J-V characteristic in the dark (broken curve) and under illumination (full curve) of an 0.25 cm$^2$ In$_2$Se$_3$ electrode in 1 mM Na$_2$SeO$_4$ electrolyte. With an illumination intensity of 80 mW/cm$^2$. .......................................................................................................................... 115
Figure IV-57. J-V characteristic in the dark (broken curve) and under illumination (full curve) of an 0.25 cm$^2$ In$_2$Se$_3$ electrode in Na$_2$S$_2$O$_3$ 0.1M electrolyte. The illumination intensity was 80 mW/cm$^2$. .......................................................................................................................... 116
Figure IV-58. Cyclic voltammetry of a 0.25 cm$^2$ In$_2$Se$_3$ electrode immersed in 1 mM Na$_2$SeO$_4$ electrolyte. ......................................................................................................................... 117
Figure IV-59. Cyclic voltammetry of a 0.25 cm$^2$ In$_2$Se$_3$ electrode immersed in Na$_2$S$_2$O$_3$ 0.1 M electrolyte. ......................................................................................................................... 118
Figure IV-60. Mott-Schottky diagram of a 1 cm$^2$ In$_2$Se$_3$ electrode. Experiment performed in a 0.1 M KCl solution at pH 7 with a frequency range modulated between 100 Hz and 1 kHz. .......... 119
Figure IV-61. J-V characteristic of an In$_2$Se$_3$ (1 cm$^2$) electrode in the dark (broken curve) and under an illumination intensity of 80 mW/cm$^2$ (full curve) in Na$_2$SeO$_4$ 1 mM electrolyte. ............ 120
Figure IV-62. Cyclic voltammetry of 1 cm$^2$ In$_2$Se$_3$ electrode in Na$_2$SeO$_4$ 1 mM electrolyte. ...... 121
Figure IV-63. Cyclic voltammetry of 1 cm$^2$ In$_2$Se$_3$ electrode in Na$_2$S$_2$O$_3$ 0.1 M electrolyte. ...... 121
Figure IV-64. X-ray diffraction pattern of Ga$_2$O$_3$ sample (° peaks correspond to $\alpha$-Ga$_2$O$_3$ and * peaks correspond to $\beta$-Ga$_2$O$_3$ after WinXPOW database). ......................................................... 124
Figure IV-65. SEM Ga$_2$O$_3$. .......................................................................................................... 125
Figure IV-66. Ga$_2$O$_3$ EDXS analysis after synthesis. ................................................................ 126
Figure IV-67. XRD of the synthesized Ga$_2$O$_3$ powder after one hour of calcination at 500°C sample (° peak correspond to $\alpha$-Ga$_2$O$_3$ and * peaks correspond to $\beta$-Ga$_2$O$_3$ after WinXPOW database). ........................................................................................................................... 126
Figure IV-68. $\beta$-Ga$_2$O$_3$ scan microscopy powder calcinated at 500°C for one hour. ................ 127
Figure IV-69. EDXS after one hour calcination at 500° C. ....................................................... 127
Figure IV-70. XRD of Ga$_2$O$_3$ calcinated at 500°C for 5 hours (° peak correspond to $\alpha$-Ga$_2$O$_3$ and * peaks correspond to $\beta$-Ga$_2$O$_3$ after WinXPOW database). ................................................. 128
Figure IV-71. STEM nanofibres of gallium oxide calcinated at 500°C for 5 hours. .................. 129
Figure IV-72. XRD pattern of $\beta$-Ga$_2$O$_3$ calcinated for 5h at 1000°C (* peaks correspond to $\beta$-Ga$_2$O$_3$ after WinXPOW database). ................................................................................................. 129

# Index

Figure IV-73. Brillouin zone structure of $\beta$-Ga$_2$O$_3$. The $k$ points that overlap in the $\Gamma$ point of the supercell are marked with different letters. ............................................................................. 130

Figure IV-74. $\beta$-Ga$_2$O$_3$ TEM pictures are presented in (A) and (B) and HRTEM in (C),(D),(E) and in (F) is presented the optical diffraction with a grid distance of d$_{011}$ 2.495 Å. ............................ 131

Figure IV-75. Reflectance function (A) and of normalized Reflectance function (B) of $\beta$-Ga$_2$O$_3$ calcinated at 1000° C (•••), anatase PC500 (—), rutile R15 (---), In$_2$Se$_3$ (•••) and Ga$_2$O$_3$ calcinated at 500° C (—) are compared. ................................................................................... 131

Figure IV-76. (A) $\beta$-Ga$_2$O$_3$ indirect band gap determination (B) $\beta$-Ga$_2$O$_3$ direct band gap determination. ............................................................................................................................... 132

Figure IV-77. Acetaldehyde degradation with 1g pressed $\beta$-Ga$_2$O$_3$ calcinated (for 5h at 1000°C) photocatalyst under 1mW/cm$^2$ UV-A light. ........................................................................... 133

Figure IV-78. Elemental electron microscopy analysis of Ga$_2$O$_3$ film (upper view) (A) and corresponding gallium mapping (B). ....................................................................................... 135

Figure IV-79. Elemental electron microscopy analysis of Ga$_2$O$_3$ film (side view) (A), gallium mapping (B), chlorine mapping (C), oxygen mapping (D) and carbon mapping (E). ......................... 135

Figure IV-80. EDXS of Ga$_2$O$_3$ film calcinated for one hour at 500°C. ................................................ 136

Figure IV-81. Mott-Schottky diagram of a 0.25 cm$^2$ Ga$_2$O$_3$ electrode. immersed in a 0.1 M KCl solution at pH 7 with a frequency range modulated between 100 Hz and 1 kHz. ................ 137

Figure IV-82. $J$-$V$ characteristic of a 0.25 cm$^2$ Ga$_2$O$_3$ electrode in the dark (broken curve) and under illumination (80 mW/cm$^2$) (full curve) immersed in a 0.1 M Na$_2$S$_2$O$_3$ electrolyte. ................. 138

Figure IV-83. $J$-$V$ characteristic of a 0.25 cm$^2$ Ga$_2$O$_3$ electrode in the dark (broken curve) and under illumination (80 mW/cm$^2$) (full curve) immersed in 1 mM Na$_2$SeO$_4$ electrolyte. ................... 139

Figure IV-84. Cyclic voltammetry of a 0.25 cm$^2$ Ga$_2$O$_3$ electrode immersed in 0.1 M Na$_2$S$_2$O$_3$ electrolyte. ............................................................................................................................... 140

Figure IV-85. Cyclic voltammetry of a 0.25 cm$^2$ Ga$_2$O$_3$ electrode immersed in 1 mM Na$_2$SeO$_4$ electrolyte. ............................................................................................................................... 140

## Index of tables

Table IV-1. Visible light intensity values under the different filters for the gas-phase set up lamp. ..... 52
Table IV-2. $NO_x$ and NO degradation results under UV light with P25 photocatalyst. ........................ 56
Table IV-3. NO degradation results under visible light using different filters with P25 photocatalyst. 57
Table IV-4. $NO_x$ degradation results under visible light using different filters with P25 photocatalyst. 57
Table IV-5. Acetaldehyde degradation results under UV-A light with P25 photocatalyst. ................... 58
Table IV-6. Acetaldehyde degradation results under visible light using different filters with P25 photocatalyst. ........................................................................................................................... 59
Table IV-7. Photonic efficiency values of the employed photocatalysts. ............................................. 73
Table IV-8. Elemental analysis of the S-doped $TiO_2$ with adsorbed $Fe^{3+}$ photocatalyst powder. ........ 77
Table IV-9. Photocatalytic efficiency results of the comparison between the different commercial photocatalysts under 30 $mW/cm^2$ UV-A light (320 nm filter). (Figures IV-30 and IV-37) ........... 90
Table IV-10. Photocatalytic efficiency results of the comparison between the different commercial photocatalysts under 0.896 $mW/cm^2$ visible light, using a filter $\lambda > 420nm$. (Figures IV-29 and IV-37) .................................................................................................................................... 91
Table IV-11. $NO_x$ degradation results under visible light using different filters with S-doped $TiO_2$-$Fe^{3+}$ photocatalyst. ........................................................................................................................ 92
Table IV-12. NO degradation results under visible light using different filters with S-doped $TiO_2$-$Fe^{3+}$ photocatalyst. ........................................................................................................................ 93
Table IV-13. Acetaldehyde degradation results under visible light using different filters with S-doped $TiO_2$-$Fe^{3+}$ photocatalyst. ........................................................................................................ 94
Table IV-14. EDXS $In_2Se_3$ results. ................................................................................................ 111
Table IV-15. Photonic efficiency result of acetaldehyde degradation under UV light with a calcinated $\beta$-$Ga_2O_3$ photocatalyst. .................................................................................................... 134

## Acknowledgements

First I would like to gratefully acknowledge the supervision and support of Professor Detlef Bahnemann during this work for all the numerous scientific discussions and help in the research area, but also for his opinions and invaluable advises.

I would like to express my sincere appreciation to all members of the Research Group of Photokatalyse & Nanotechnologie Detlef, Anja, Marco, Irina, Daniel, Astrid, Ralf, Julia, Olga, Tarek, Hanan, Amer, Razan, Jonathan, Clarissa, Elias, Nadia, Cecilia, Aminee, Anna and extend this feeling to the entire Institut für Technische Chemie (TCI) specially to Professor Thomas Scheper and his group for their assistance and the very good atmosphere of work.

All my gratitude to Professor Teruhisa Ohno for the collaboration project; to Viktor Yarovyi, Frank Steinbach and Stella Kittel for the TEM analysis.

To the DAAD (Deutsche Akademische Austausch Dienst) thanks for the scholarship.

To my whole family and friends especially to my parents Victor Manuel Menéndez Trejo and Maria Teresa Irma Flores de Menéndez, thank you for believing in me and standing by me along all this time. You are part of my soul and the reason of my life.

To Marcelino Menéndez Pérez and Ernesto Flores Ayala my grandfathers to whom I dedicate this work. I can´t see You because You aren't here anymore, but I can feel your heart near mine.

*Ahí,*
*en el silencio de la sala,*
*brillaba el bisturí,*
*como si fuera una estrella prisionera,*
*en la mano sabia del Cirujano.*

*Da,*
*in dem stillen Saal,*
*leuchtet das Skalpell*
*als ob´s ein gefangener Stern wäre,*
*in der starken weisen Hand des Chirugs.*

# Symbols and abbreviations

| Symbol | Description | Units |
|---|---|---|
| $A$ | area | [cm$^2$] |
| $A$ | electron affinity | |
| $c$ | light velocity | [m*s$^{-1}$] |
| $C_o$ | initial$_{DCA}$ concentration | [mol*L$^{-1}$] |
| $C_{sc}$ | space charge capacity | |
| $C_{ox}$ | concentration of the $Ox$ species | |
| $C_{red}$ | concentration of the $Red$ species | |
| $d$ | thickness of the semiconductor | |
| $d_H$ | distance | |
| $d_{sc}$ | semiconductor distance | |
| $D$ | diffusion coefficient | |
| $D$ | electron donor | |
| $D_{red}$ | density of the electronic state for the occupied state | |
| $D_{ox}$ | density of the electronic state for the empty state | |
| DCA | dichloroacetic acid | |
| $E_c(x)$ | energy of the lower edge of the conduction band | |
| $E_c^b$ | energy level of the conduction band in the bulk | |
| $E_F$ | Fermi level energy | |
| $E_{F,redox}^0$ | standard electrochemical potential of the redox couple | |
| $E_{red}^0$ | reduction energy | |
| $E_{ox}^0$ | oxidation energy | |
| $E_g$ | band gap | |
| $E_v(x)$ | energy of the upper edge of the valence band | |
| $E_v^b$ | energy level of the valence band in the bulk | |
| $E_v^s$ | energy of the valence band at the surface of the semiconductor | |
| $e$ | electron charge | 1.602*10$^{-19}$ [C] |
| eV | electron Volt | |
| $f(E)$ | Fermi energy distribution function | |
| $f(E_c)$ | conduction band energy function | |
| $h$ | Planck constant | 6.63*10$^{-34}$ [J*s] |
| $I$ | light intensity | [J*s$^{-1}$cm$^{-2}$] |
| $I_0$ | incident photon flux | |
| $j_c^-$ | cathodic current | |
| $j_c^+$ | anodic current | |
| $j_g$ | the generation current | |
| $j_{ph}$ | photocurrent density | |
| $J_0$ | light flux | [mol* s$^{-1}$cm$^{-2}$] |
| $j_{diff}$ | diffusion current density | |
| $j_{sc}$ | semiconductor current density | |
| $j_0$ | saturation current density | |

## Symbols and abbreviations

| | | |
|---|---|---|
| $j_v^0$ | saturation current density | |
| $k_0$ | electron transfer rate constant | |
| $k$ | initial rate constant | [s⁻¹] |
| $L_p$ | hole diffusion length | |
| $m_e^*$ | effective mass of electrons | |
| $m_h^*$ | effective mass of holes | |
| $m_0$ | electron mass in vacuum | |
| $m_e$ | electron rest mass | 9.1096·10⁻³¹ [kg] |
| $n_0$ | charge carrier density of electrons in the bulk | |
| $n_s$ | density of free electrons on the surface | |
| $N_d$ | ionized donor density | |
| $N_a$ | ionized acceptor density | |
| $n(x)$ | electron density | |
| $n_i$ | intrinsic carrier density | |
| $N_A$ | Avogadro's number | 6.22*10²³ [mol⁻¹] |
| $n_i$ | intrinsic electron density | |
| $N_c$ | density of states at the lower edge of the conduction band | |
| $N_v$ | density of states at the upper edge of the valence band | |
| $N_D$ | density of donor states | |
| $p(x)$ | hole density | |
| $p_0$ | equilibrium hole density | |
| $p_d$ | hole density | |
| $p_s$ | hole density on the semiconductor surface | |
| $p_0$ | charge carrier density of holes in the bulk | |
| $T$ | temperature | [°C] |
| UV | ultra violet | |
| $V$ | volume | [L] |
| $W_{red}(E)$ | distribution of the occupied electronic level | |
| $W_{ox}(E)$ | distribution of the empty level | |
| $x$ | distance between surface and acceptor or donor | |
| $\alpha$ | depth of penetration (of light) | |
| $\beta$ | radian measurement | |
| $\Delta\phi_H$ | potential drop | |
| $\Delta\phi_{sc}$ | potential drop across space charge region | |
| $\Delta\phi_{sc}^0$ | overpotential created upon illumination | |
| $\Delta E_F$ | Fermi level energy difference | |
| $\varepsilon_0$ | dielectric constant | 8.854·10⁻¹² [F m⁻¹] |
| $\varepsilon$ | permittivity of free space | |
| $\theta$ | Bragg angle | degree |
| $\lambda$ | wavelength | [nm] |
| $\xi$ | photonic efficiency | [%] |
| $\rho(x)$ | density | |

# Symbols and abbreviations

| | |
|---|---|
| $\rho(E_c)$ | distribution of energy states via conduction band |
| $\rho(E)$ | distribution of energy states |
| $\tau_{tr}$ | average transit time |
| $\Phi$ | quantum yield |

## Kurzfassung

Diese Arbeit beschäftigt sich mit der Entwicklung von in der Photokatalyse verwendeten Nanopartikeln, in Abhängigkeit von deren Stabilität bei dem Abbau von Schadstoffen in der wässrigen Phase und / oder in der Gasphase. Verschiedene Parameter, welche sich an der spezifischen Anwendung jedes Photokatalysators orientieren, wurden getestet. So wurde die Stabilität von Ag-TiO$_2$, die Verwendung von S-TiO$_2$-Fe$^{3+}$ unter Einstrahlung von sichtbarem Licht und die Synthese von zwei Photokatalysatoren mit schmaler und breiter Bandlücke wie In$_2$Se$_3$ und β-Ga$_2$O$_3$ für photochemischen Abbau in der Gasphase untersucht.

Die Stabilität des Photokatalysators Ag-TiO$_2$ wurde mittels des photokatalytischen Abbaus von Dichloressigsäure (DCA) als Funktion der Recycling untersucht. Die photokatalytische Aktivität wurde durch Messungen der Rate, mit welcher H$^+$-Ionen während des photokatalytischen Abbaus von DCA freigesetzt wurden, bestimmt. Bestätigt wurden diese Ergebnisse durch Bestimmung des entfernten Kohlenstoff (total organic carbon). Die photokatalytischen Abbaureaktionen wurden bei den pH-Werten 3 und 10 für eine Versuchsreihe mit Ag-TiO$_2$ Photokatalysatoren durchgeführt. Alle Ag-TiO$_2$ und die reinen TiO$_2$ Photokatalysatoren zeigten eine Abnahme der photokatalytischen Aktivität bei Durchführung der Photoabbau-Reaktion von DCA. Die Abnahme der Aktivität kann als Vergiftung von aktiven Zentren durch Cl$^-$-Anionen während des photokatalytischen DCA Abbaus angesehen werden.

Ein mit Schwefel dotiertes Titandioxid-Material mit adsorbiertem Eisenoxid wurde gründlich als Photokatalysator in Gegenwart von unterschiedlichen Modelsubstanzen analysiert. Hierbei wurde Dichloracetsäure (DCA) in der wässriger Phase und NO$_x$ oder Acethaldehyd in der Gasphase verwendet. Die Stabilität des Photokatalysators wurde als Funktion der Recycling durch den Abbau von DCA bei den pH-Werten 3,7 und 9 untersucht. Dieses Material zeigt echte photokatalytische Aktivität bei einem pH-Wert von 3 unter Verwendung von sichtbarem Licht und ist somit für zukünftige Anwendungen sehr interessant. In$_2$Se$_3$ und β-Ga$_2$O$_3$, typische Solarzellen-Materialien, wurden synthetisiert und untersucht, da ihre Bandlücken-Eigenschaften diese für die Leistung von unterschiedlichen Anwendungen interessant machen. Sie wurden mit traditionellen TiO$_2$ Halbleitermaterialien verglichen, indem die charakteristischen Tests, wie der Abbau von Acetaldehyd oder NO$_x$ für Gasphasenreaktionen durchgeführt wurden.

**Stichworte**: Photokatalyse, Photoabbau-Reaktion, Titandioxid, Dotierung, Stabilität.

## Abstract

This study has been focused on the development of semiconductor nanoparticles applied in photocatalysis depending on their stability for the decomposition of pollutants in aqueous and/or gas phase. Different parameters were tested related to their specific application of each photocatalyst like the stability of Ag-TiO$_2$, the use of S-TiO$_2$-Fe$^{3+}$ under visible light and the synthesis of two photocatalysts with a short and a wide band gap, such as In$_2$Se$_3$ and $\beta$-Ga$_2$O$_3$ respectively, for photochemical gas decomposition The stability of Ag-TiO$_2$ photocatalysts was examined for the photocatalytic degradation of dichloroacetic acid (DCA) as a function of the recycling times. The photocatalytic activity was investigated by measuring the rate of H$^+$ ions released during the photodegradation of DCA and confirmed by measuring the total organic carbon removal. The photodegradation reactions were studied at pH 3 and pH 10 for a series of Ag-TiO$_2$ photocatalysts. All the Ag-TiO$_2$ and bare TiO$_2$ photocatalysts showed a decrease in photocatalytic activity on recycling for the DCA photodegradation reaction. The decrease in activity can be attributed to poisoning of active sites by Cl$^-$ anions formed during the photocatalytic DCA degradation. The photocatalytic activity was, however, easily recovered by a simple washing technique. The reversibility of the poisoning (presumably by the chloride ions) is taken as evident to support the idea that the recycling of Ag-P25 TiO$_2$ photocatalysts does not have a permanent negative effect on their photocatalytic performance for the degradation of DCA. The choice of the preparation procedure for the Ag-TiO$_2$ photocatalysts is shown to be of significant importance. A sulfur doped titanium dioxide material adsorbed with ferric oxide, was rigorously analyzed as a photocatalyst in presence of different model compounds; dichloroacetic acid (DCA) in a liquid phase and NO$_x$ or acetaldehyde in a gas phase. The stability of the photocatalyst was examined as a function of the recycled times by the degradation of DCA at the different pH conditions pH 3, 7 and 9. This material presents real photoactivity at pH 3 under visible light, able to be considered for further commercial applications.

In$_2$Se$_3$ and $\beta$-Ga$_2$O$_3$ common solar cell semiconductor materials were synthesized and studied because of its band gap properties for the performance of different applications, comparing the traditional TiO$_2$ semiconductor characteristics, as a photocatalytic material, by the degradation of acetaldehyde or NO$_x$ pollutants for gas phase reactions.

**Key words**: Photocatalysis, semiconductor, photodeposition, doping, stability, visible-light.

## I. Introduction and Problem Statement

The earth is facing difficult problems regarding the global environment since photocatalysis involves important science for creating new technology with energy resources from sunshine in order to clean the environment applying the possibility of solar conversion energy by semiconductors or sensitizers. Catalysis under light irradiation is called photocatalysis. Some attempts were made in the past to define the term photocatalysis. Indeed one of the IUPAC Commissions defined photocatalysis as a catalytic reaction involving light absorption by a catalyst or a substrate [1-3].

Catalysis is the action of a catalyst, which is a substance that increases the rate of reaction without modifying the overall standard Gibbs energy change in the reaction. Catalysis refers simply to a process in which a substance (the catalyst) accelerates, through intimate interactions with the reactant(s) and concomitantly providing a lower energy pathway, an otherwise thermodynamically favored but kinetically slow reaction, with the catalyst fully regenerated quantitatively at the conclusion of the catalytic cycle. When photons are also involved, the expression photocatalysis can be used to describe, this process without the implication of any specific mechanism, as the acceleration of a photoreaction by the presence of a catalyst. The catalyst may accelerate the photoreaction by interacting with the substrate(s) either in its ground state, in its excited state or with the primary product (of the catalyst), depending on the mechanism of the photoreaction [4]. The role of light is to form the active (excited) state of the catalyst or to produce more active sites on its surface (for a heterogeneous system) during photoexcitation. But if during the photocatalytic process the reactant molecule is strongly (chemically) bonded to surface atoms, the photocatalytic site becomes inactive and said to be poisoned.

The photoactivity of the photocatalyst $TiO_2$ (anatase and rutile structure) under UV light is well known, however, this technique still has limited applications due to the relatively low photoactivity under visible light. There are different methods to increase the photoactivity of a photocatalyst, like the photodeposition of noble metals on the surface of the semiconductor, by doping or by photosensitization.

# Introduction

Here the synthesis of different photocatalysts and their stability, which were tested to check the durability and performance of the developed material after photocatalytic experiments, are introduced. Specifically the photodeposited Ag-$TiO_2$, S-doped $TiO_2$ nanoparticles and semiconductors with extreme short and wide band gaps as $In_2Se_3$ or $\beta$-$Ga_2O_3$ respectively, were tested by diverse photocatalytic experiments. The photoactivity of these nanomaterials was related to the photonic efficiency ($\xi$) term, which had been suggested [4] to describe the number of transformed product molecules formed divided by the number of photons at a given wavelength incident on the reactor cell (flat parallel windows). Alternatively, the photonic efficiency can also be described by relating the initial rate of the event to the rate of incident photons reaching the reactor.

The intention of this work is to analyze different semiconductor systems by photocatalytic reactions and determine the specific driving force for each one to increase the photonic efficiency by diverse methods for further possible applications under the visible range. This dissertation covers the determination of photocatalytic durability for deposited Ag-$TiO_2$ particles; proves out the photoactivity of S-$TiO_2$ material under visible light and develops synthesis of materials like $In_2Se_3$ or $\beta$-$Ga_2O_3$ to be evaluated as photocatalysts.

The method of photodeposition was analyzed on the $TiO_2$ photocatalyst surface which was modified by deposition of metallic silver loadings under UV-A light. The influences of the synthesis and reaction conditions were examined on the photoactivity of the photodeposited Ag-$TiO_2$ by decomposing dichloroacetic acid (DCA) in aqueous phase.

Many attempts have been tried along the last years to modify the $TiO_2$ because it has a relative wide band gap (ca. 3 eV). Until now the modifying efforts haven't shown any photocatalysts which could be applied due to their instability in the photocatalytic systems. Therefore the S-doped $TiO_2$-$Fe^{3+}$ photocatalyst was developed in order to increase the range of solar spectrum use. In consequence several photoactivity experiments were realized with S-doped-$TiO_2$-$Fe^{3+}$ material in aqueous and gas phases. Different pH conditions should determine the stability and efficiency of the photocatalyst tested as a visible photoactive material by decomposing DCA in the aqueous phase. In the gas phase $NO_x$ and acetaldehyde were also model pollutants to be decomposed under visible light irradiation.

## Introduction

$In_2Se_3$ is an interesting material used commonly in solar cells, therefore was synthesized and characterized. This semiconductor has a thin band gap but is not stable in aqueous phase, therefore photocatalytic experiments in gas phase were suggested to show any possible application under visible light of this material.

$\beta$-$Ga_2O_3$ is a material recently used for water splitting with the intention of obtaining hydrogen, which starts beeing a very important fuel for the near future. Because of this was synthesized as a p-type semiconductor and characterized. Properties like its wide band gap and high stability have been considered for the performance of photocatalytic experiments. The results were compared to the traditional $TiO_2$ semiconductor characteristics for possible applications as a photocatalytical material.

## II. Theory

### II.1. Fundamentals

All semiconductors posess, a forbidden energy region or band gap in which energy states cannot exist. Energy bands are only permitted above and below this energy gap. Figure II-1 shows band gap energies and band gap edges of the most frequently used photocatalysts. The upper bands are called the conduction bands, the lower ones the valence bands. The band structure is a function of the three-dimensional wave vector within the Brillouin zone. If both, the conduction band minimum and the valence band maximum occur at the center of the Brillouin zone and at the same vector, the energy difference, $E_g$, is a so called direct band gap. But if the lowest minimum of the conduction band and the maximum of the valence band occur at a different wave vector, $E_g$ is termed an indirect band gap.

Titanium dioxide has semiconducting properties which makes it an attractive material to be used as a photoactive catalyst. $TiO_2$ is widely used for air purification, deodorization, sterilization, anti-fouling and mist removal under UV light [5]. Activity of $TiO_2$ depends on its surface area, porosity and acid-basic properties. It was also found that the photoactivity depends on the crystallite size and the relative abundance of the crystallite phases (anatase/rutile). Both, the crystallite size and crystalline phases, modify the $TiO_2$ band gap. The pristine $TiO_2$ is only active upon ultraviolet light ($\lambda < 351$ nm) because of its band gap (3.2 eV in the anatase $TiO_2$ crystalline phase).

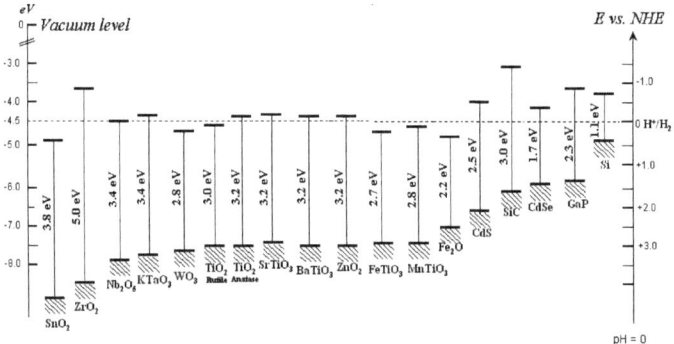

Figure II-1. Position of energy bands at the surface of various semiconductors in aqueous electrolytes at pH 0 [6] (modified).

To improve the photocatalytic reactivity of $TiO_2$ and to extend its light absorption into the visible region, several attempts have been made: one is to dope transition metals into $TiO_2$ [7], and another is to form reduced $TiO_x$ photocatalysts [8].

Both approaches introduce impurity/defect states in the band gap of $TiO_2$, which leads to a visible light absorption of $TiO_2$. However, transition metal doping, where quite localized d states appear deep in the band gap of the host semiconductor, often results in the increase of the carrier recombination. Therefore, the lifetime of the mobile carriers may become shorter, giving lower photocatalytic activity. Reducing $TiO_2$ introduces localized oxygen vacancy states located at 0.75-1.18 eV below the conduction band of $TiO_2$. These vacancies may promote photoreduction activity because a redox potential of the hydrogen evolution ($H_2/H_2O$) is located just below the conduction band of $TiO_2$. In 2001, Asahi et al. [9], presented a new type of visible light sensitive photocatalyst nitrogen-doped $TiO_2$. Since Asahi's paper, other non-metal doped $TiO_2$ photocatalysts were reported. $TiO_2$ doped with fluorine [10], iodine [11] and phosphor [12] showed higher photocatalytic activity under UV light and $TiO_2$ doped with nitrogen [13,14], carbon [15], sulfur [16] and codoped with nickel and boron [17] showed high photocatalytic activity under visible light. Sato et al.,[18] have shown efficient photooxidation of CO under visible irradiation by a nitrogen-doped $TiO_2$. C-doped $TiO_2$ was obtained by an acid-catalyzed sol-gel process [19] or by the oxidative annealing of titanium carbide [20].

AOPs are the advanced oxidative degradation processes for the organic compounds, dissolved or dispersed in aquatic media by catalytic, chemical and photochemical methods [21]. These processes rely on the generation of organic radicals, which are produced either by photolysis of organic substrate or by reaction with hydroxyl radicals. Dissolved molecular oxygen and peroxides trap these radicals for complete mineralization of organics. AOPs are categorized into homogeneous and heterogeneous processes depending on the physical state of the catalyst.

Recently, heterogeneous and homogeneous photocatalytic detoxification methods ($TiO_2/H_2O_2$, $Fe^{3+}/H_2O_2$) have shown great promise in the treatment of industrial wastewater, groundwater and contaminated air [22]. It is well established that by the irradiation of an aqueous $TiO_2$ suspension with light energy greater than the band gap energy of the semiconductor ($E_g > 3.2$ eV) conduction band electrons ($e^-$) and valence band holes ($h^+$) are generated (Figure II-2). Part of the photogenerated carriers recombine in the bulk of the semiconductor, while the rest reach the surface, where the holes, as well as the electrons act as powerful oxidants and reductants respectively. The photogenerated electrons react with the adsorbed molecular $O_2$ on the Ti(III)-sites, reducing it to a superoxide radical anion $O_2^{\bullet-}$, while the photogenerated holes can oxidize either the organic molecules directly, or the $OH^-$ ions and the $H_2O$ molecules adsorbed at the $TiO_2$ surface to $OH^{\bullet}$ radicals. These radicals together with other highly oxidant species (e.g. peroxide radicals) are reported to be responsible for the primary oxidizing step in photocatalysis. The $OH^{\bullet}$ radicals formed on the illuminated semiconductor surface are very oxidizing agents, with a standard oxidation potential of 2.8 V (vs. NHE at pH 0) [23]. These can easily attack the adsorbed organic molecules or those located close to the surface of the catalyst, thus leading finally to their complete mineralization.

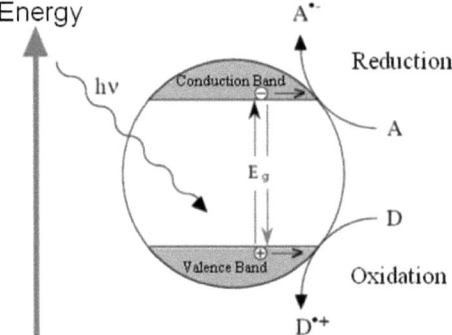

Figure II-2. Principle mechanism of photocatalysis [24].

## II.1.1. Semiconductor-interface behavior in absence of redox systems

Photocatalytic reactions in aqueous phase studied in the semiconductor-electrolyte interface in absence of redox species explained, that when the semiconductor is brought into contact with an electrolyte, the so called Helmholtz double layer, a potential drop of $\Delta\phi_H$ over a distance $d_H$ is formed on the solution side.

On the solid side, the counter charge (i.e., the charge to neutralize that on the solution side due to bonding or specific adsorption of, for example, hydroxyl ions) is distributed over a certain range below the surface. Correspondingly, a potential drop $\Delta\phi_{sc}$ is formed across the so called space charge

# Theory

region, the dimension of which is denoted as $d_{sc}$, as indicated in Figure II-3. Here the diffuse layer in the solution has been neglected assuming that a high ion concentration is applied. The potential and charge distribution within the space charge region in the solid is described by the Poisson equation as follows [6]:

$$\frac{d^2 \Delta \phi_{sc}}{dx^2} = -\frac{1}{\varepsilon \varepsilon_0} \rho(x) \tag{II.1}$$

in which $\varepsilon$ is the dielectric constant of the material and $\varepsilon_0$ is the permittivity of free space. The charge density $\rho(x)$ is given by

$$\rho(x) = e[N_d - N_a - n(x) + p(x)] \tag{II.2}$$

where $x$ is the distance from the surface to $N_d$ or $N_a$ which are the fixed ionized donor and acceptor densities, respectively, introduced by the doping of the semiconductor. The electron and hole densities, $n(x)$ and $p(x)$, vary with the distance $x$ according to the following equations

$$n(x) = N_c \exp\left(-\frac{E_c(x) - E_F}{kT}\right) \tag{II.3}$$

$$p(x) = N_v \exp\left(-\frac{E_v(x) - E_F}{kT}\right) \tag{II.4}$$

where $N_c$ is the density of states at the lower edge of the conduction band, and $N_v$ is the density of states at the upper edge of the valence band. $E_c(x)$ is the energy of the lower edge of the conduction band which varies with the distance and $E_v(x)$ the energy of the upper edge of the valence band. $E_F$ is the Fermi level energy of the semiconductor, which is defined in solid state physics as the highest occupied energy level within the crystal at the temperature of absolute zero. For an intrinsic semiconductor, the position of its Fermi level can be calculated from

$$E_F = \frac{E_c + E_v}{2} + \frac{kT}{2} \ln\left(\frac{N_v}{N_c}\right) = \frac{E_c + E_v}{2} + \frac{kT}{2} \ln\left(\frac{m_h^*}{m_e^*}\right) \tag{II.5}$$

in which $m_e^*$ and $m_h^*$ are the effective masses of electrons and holes, respectively.

Theory

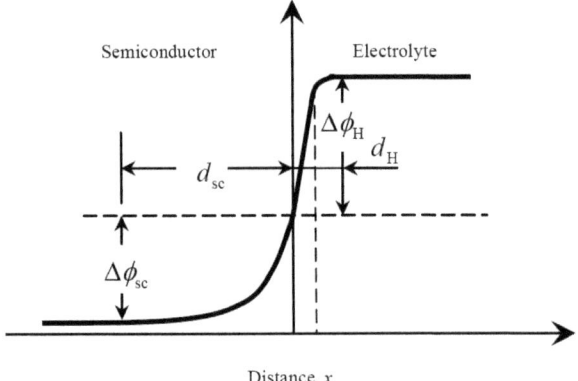

Figure II-3. Potential distribution at the semiconductor-electrolyte interface [6].

Within the space charge region, since the Fermi level is expected to be constant, the position of the energy bands $E_c(x)$ and $E_v(x)$ vary with distance. If the charge carrier densities in the bulk of the semiconductor are $n_0$ for electrons and $p_0$ for holes, then

$$n(x) = N_0 \exp\left(-\frac{E_c(x) - E_c^b}{kT}\right) = N_0 \exp\left(-\frac{e\Delta\phi_{sc}(x)}{kT}\right) \quad (\text{II.6})$$

$$p(x) = p_0 \exp\left(-\frac{E_v(x) - E_v^b}{kT}\right) = p_0 \exp\left(-\frac{e\Delta\phi_{sc}(x)}{kT}\right) \quad (\text{II.7})$$

in which $E_c^b$ and $E_v^b$ are the energy levels of the conduction and valence bands in the bulk. These two equations imply a Boltzmann distribution of charge carriers in the space charge region. In the bulk of the semiconductor, i.e., at sufficient distance from the surface, charge neutrality must be obeyed. Then,

$$N_d - N_a = n_0 - p_0 \quad (\text{II.8})$$

Inserting Equations (II.2), (II.6), (II.7) and (II.8) into Eq. (II.1) one obtains

$$\frac{d^2\Delta\phi_{sc}}{dx^2} = -\frac{e}{\varepsilon_0}\left[n_0 - p_0 - n_0 \exp\left(-\frac{e\Delta\phi_{sc}}{kT}\right) + p_0 \exp\left(\frac{e\Delta\phi_{sc}}{kT}\right)\right] \quad (\text{II.9})$$

# Theory

Without going into much mathematical detail, the capacity of the space charge region is derived by solving the Poisson-Boltzmann equation as

$$C_{sc} = \frac{\varepsilon_0}{L_D} \cosh\left(\frac{e\Delta\phi_{sc}}{kT}\right) \qquad (II.10)$$

in which the so called Debye length $L_D$ is defined by

$$L_D = \left(\frac{\varepsilon_0 kT}{2n_i e^2}\right)^{1/2} \qquad (II.11)$$

where $n_i$ is the intrinsic carrier density given by the equilibrium equation $n_0 p_0 = n_i^2$. From equation (II.10) one recognizes that the space charge capacity $C_{sc}$ depends strongly on the potential drop $\Delta\phi_{sc}$ across the space charge region.

## II.1.2. Semiconductor-interface behavior in presence of redox systems

In the presence of a redox system which is dissolved in the electrolyte, an energy difference exists between the Fermi level of the semiconductor and the redox couple. To reach the equilibrium conditions charge carrier transfer occurs across the semiconductor-liquid interface via the energy bands, i.e., the conduction or valence band of the semiconductor. At the equilibrium point, the Fermi level of the redox system, $E_{F,\,redox}$ is equivalent to the electrochemical potential of the electrons in the redox system, as shown in the following equation:

$$E_{F,\,redox} = \overline{\mu}_{e,\,redox} = \mu^0_{e,\,redox} + kT\ln\left(\frac{C_{ox}}{C_{red}}\right) \qquad (II.12)$$

in which $C_{ox}$ and $C_{red}$ are the concentrations of the Ox and Red species from the following reaction:

$$Ox + e^- \longleftrightarrow Red \qquad (II.13)$$

Conventionally, the corresponding redox potential is given on a scale using the normal hydrogen electrode (NHE) or the saturated calomel electrode (SCE) as reference electrode. The Gerischer model [25], which has the advantage that energy levels of localized electron states in an electrolyte solution can be introduced, is usually employed to illustrate the charge carrier transfer between the semiconductor and the redox system at the interface. In a very simplified version of the Gerischer model as illustrated in Figure II-4, the redox system is characterized by a set of occupied states centered around the energy value of $E_{red}^0$ and a set of empty states centered around $E_{ox}^0$. Correspondingly, the densities of the electronic states, i.e., $D_{red}$ for the occupied and $D_{ox}$ for the empty states, are proportional to the concentration of the reduced ($C_{red}$) and oxidized ($C_{ox}$) species in the redox system, respectively, as given by

$$D_{red}(E) = C_{red} W_{red}(E) \tag{II.14}$$

$$D_{ox}(E) = C_{ox} W_{ox}(E) \tag{II.15}$$

in which $W_{red}(E)$ describes the distribution of the occupied electronic level and $W_{ox}(E)$ the distribution of the empty level. These two distribution functions are dependent on the reorganization energy $\lambda$ (Figure II-4a). This is accompanied by the rearrangement of the solvation shell and the solvent dipoles after the fast electron transfer between the reduced and oxidized species in the solution, and have the following forms:

$$W_{red}(E) = (4kT\lambda)^{-1/2} \exp\left[-\frac{(E - E_{F,redox}^0 - \lambda)^2}{4kT\lambda}\right] \tag{II.16}$$

$$W_{ox}(E) = (4kT\lambda)^{-1/2} \exp\left[-\frac{(E - E_{F,redox}^0 + \lambda)^2}{4kT\lambda}\right] \tag{II.17}$$

As shown in Figure II-4b, at equal concentrations, the distributions of the reduced and oxidized species are equivalent at the standard electrochemical potential of the redox couple, $E_{F,redox}^0$. By varying the concentrations of the redox species, one can shift the Fermi level (where $D_{red}$ and $D_{ox}$ are equal) of the redox system according to the Nernst equation (Figure II-4c).

# Theory

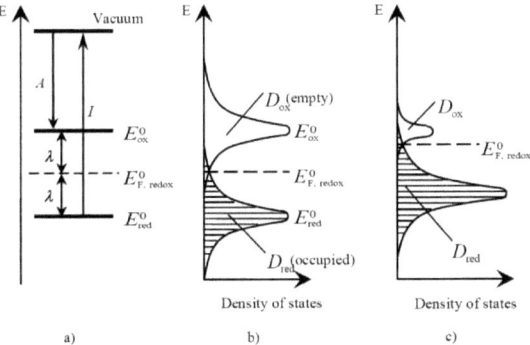

Figure II-4. Electron energies of a redox system vs. density of states: a) $E_{red}^0$ for occupied states, $E_{ox}^0$ for empty states, $A$ as electron affinity and $I$ as ionization energy of the redox system; b) the corresponding distribution functions at $C_{ox}=C_{red}$; c) the distribution functions at $C_{ox}<<C_{red}$ [6].

At the semiconductor-electrolyte interface, the electron transfer rate depends on the density of energy states on both sides of the interface. For example, the electron transfer from the redox system to the conduction band of the semiconductor generates an anodic current, described as followed

$$j_c^+ = ek_0 C_{red} \int_{E_F}^{\infty} (1-f(E))\rho(E) \exp\left[-\frac{(E-E_{F,redox}^0 - \lambda)^2}{4kT\lambda}\right] dE \qquad (II.18)$$

The subscript "c" and superscript "+" of $j$ indicate that the current is generated via the "conduction" band and is an "anodic" current. $k_0$ is the electron transfer rate constant. $\rho(E)$ is the distribution of energy states in the semiconductor. $f(E)$ is the Fermi energy distribution function as given in

$$f(E) = \frac{1}{1+\exp\left(\dfrac{E-E_F}{kT}\right)} \qquad (II.19)$$

In most cases, the overlap between energy states on both sides of the interface is limited to a small energy range, the electron transfer is therefore assumed to be within 1 $kT$ at the edge of the conduction band. The integral in Eq. (II.18) can be approximated using $dE=1$ $kT$ and $E = E_c^s$. Then one obtains

$$j_c^+ = ek_0(1-f(E_c))\rho(E_c)\left\{C_{red}\exp\left[-\frac{\left(E_c^s - E_{F,redox}^0 - \lambda\right)^2}{4kT\lambda}\right]\right\} \quad \text{(II.20)}$$

in which the product of $C_{red}$ and the exponential function corresponds to the density of occupied states of the redox system at the energy of the lower edge of the conduction band on the surface where $E = E_c^s$. The density of states in the semiconductor at the lower edge of the conduction band is

$$\rho(E_c) = N_c = \frac{2(2\pi m_c^* kT)^{3/2}}{h^3} \quad \text{(II.21)}$$

where $N_c$ is the density of energy states within few $kT$ above the conduction band edge. Usually, most of the energy states in the conduction band are empty, i.e., $1 - f \approx 1$, therefore

$$(1 - f(E_c))\rho(E_c) \approx N_c \quad \text{(II.22)}$$

Then equation (II.20) can be written as

$$j_c^+ = ek_0 N_c C_{red}\exp\left[-\frac{\left(E_c^s - E_{F,redox}^0 - \lambda\right)^2}{4kT\lambda}\right] \quad \text{(II.23)}$$

It is obvious from Eq. (II.23) that the anodic current is independent of the electrode potential, since the equation contains only constant parameters for a given system.

The cathodic current, due to electron transfer from the conduction band of the semiconductor to the empty states of the redox system, is given by

$$j_c^- = ek_0 f(E_c)\rho(E_c) C_{ox}\exp\left[-\frac{\left(E_c^s - E_{F,redox}^0 + \lambda\right)^2}{4kT\lambda}\right] \quad \text{(II.24)}$$

in which the product $f(E_c)\rho(E_c)$ at $E = E_c^s$ is the density of occupied states at the bottom of the conduction band on the surface. It is equivalent to the density of free electrons on the surface $n_s$, which is dependent on the potential drop across the space charge region as given by

$$n_s = n_0 \exp\left(-\frac{e\Delta\phi_{sc}}{kT}\right) \quad \text{(II.25)}$$

# Theory

The net current due to the electron transfer via the conduction band is then the sum of $j_c^-$ and $j_c^+$, which continues to be non-zero until the electrochemical potentials of the electrons are equal on both sides of the interface. This leads to a further modification of the space charge region in the semiconductor, on top of the simple semiconductor-electrolyte case in the absence of redox systems. Accordingly, if one visualizes the electron energy states in the space charge region, the energy bands are usually drawn as energy vs. distance plot. The so called "band bending" simply refers to the bending or curvature in such plots, as shown in Figure II-5. Other than via the energy bands, charge transfer can also occur via the surface states at the interface, which will not be further discussed in this section.

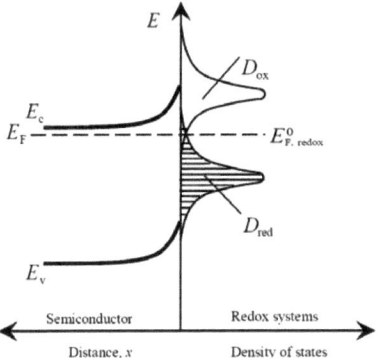

Figure II-5. Electron energies of a semiconductor electrode in contact with a redox system [6].

## II.1.3. Behavior of illuminated semiconductor-interface

If the semiconductor is illuminated by photons more energetic than its band gap energy, the equilibrium achieved at the interface in the dark is now disturbed by the excitation of electrons from the valence band to the conduction band. Besides this band-band transition, in a real, i.e., non-ideal system, there are also other electronic transitions upon light excitation, including the excitation of an electron from a donor state or an impurity level into the conduction band, and the formation of an exciton. The latter represents a bound state between an electron and a hole due to their Coulomb attraction. Its energy is close to the conduction band, and thermal excitation can hence feasibly split the exciton into an independent electron and hole. In this section, however, only the band-band transition is considered during the analysis of charge transfer at the semiconductor-liquid interface under illumination.

The absorption of a photon results in the formation of an electron-hole pair. This process is of special interest if minority carriers, i.e. the holes in a p-type and the electrons in a n-type semiconductor, are involved in the charge transfer. Thereby, the density of the minority carriers can be increased upon light excitation, by orders of magnitude, as compared with that in the dark. For an n-type semiconductor, the minority carriers (holes) generated in the space charge region are subsequently driven to the interface by the "built-in" electric field. Those generated outside this region can only reach the space charge region by diffusion, and then continue the relayed transfer driven by the electric field. A general minority carrier transfer process across the semiconductor-solution interface under illumination is illustrated in Figure II-6.

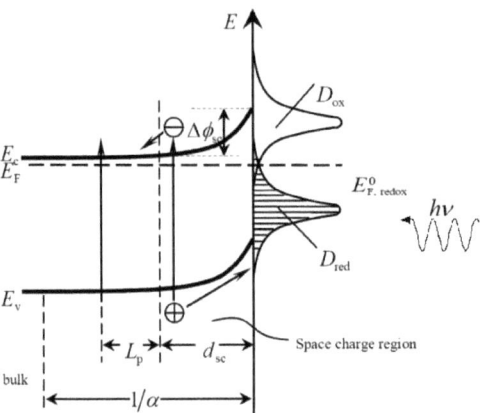

Figure II-6. Charge transfer at the semiconductor-solution interface under illumination [6].

Overall, only the holes photogenerated within the range of $L_p + d_{sc}$ will reach the surface, i.e. the interface between the semiconductor and the electrolyte, if negligible recombination within the space charge region is assumed. The distance $L_p$, termed as the hole diffusion length across, which the minority carriers travel without recombination, is usually limited due to the electron-hole recombination outside the space charge region. To obtain the photocurrent density, one needs to solve the diffusion equation for holes in the bulk using Reichman's method, as given by

$$D\frac{d^2p}{dx^2} - \frac{p-p_0}{\tau} + I\alpha\exp(-\alpha x) = 0 \qquad (\text{II.26})$$

in which $D$ is the diffusion coefficient; $p$ the hole density; $p_0$ the equilibrium hole density; the lifetime of holes; the monochromatic photon flux across the semiconductor which is dependent on the distance from the interface; the absorption coefficient of the semiconductor, a wavelength dependent parameter defined as

$$\alpha = \frac{1}{d}\ln\frac{I_0}{I} \qquad (\text{II.27})$$

where $d$ is the thickness of the semiconductor; $I_0$ the incident photon flux on the surface at $x=0$; $I$ the transmitted photon flux. Subject to the wide range of band gap energies of different semiconductors, their absorption coefficients vary typically between $10^3$ and $10^6$ cm$^{-1}$, depending on the wavelength of the incident photons. The inverse of $\alpha$ is called the depth of penetration (of light), the distance at which the radiant power decreases to 1/e of its incident value if the Naperian absorption coefficient is adopted. Accordingly, the depth of penetration of different semiconductors also varies over a wide range, typically from tens of nanometers to microns.

The three terms on the left side of Eq. (II.26) account, respectively, for the diffusion, recombination and generation of holes under illumination. Without advancing into mathematical details, one obtains the hole current density at the interface between the bulk and the space charge region, i.e., the diffusion current density $j_{diff}$, as given by

$$j_{\text{diff}} = -j_0\left(\frac{p_d}{p_0} - 1\right) + \frac{eI_0\alpha L_p}{1+\alpha L_p}\exp(-\alpha d_{sc}) \qquad (\text{II.28})$$

in which $p_0$ is the hole density in the bulk ($x=\infty$); $p_d$ is the hole density at $x=d_{sc}$ that is related to the hole density $p_s$ on the semiconductor surface under illumination via

$$p_s = p_d \exp\left(\frac{e\Delta\phi_{sc}}{kT}\right) \qquad (\text{II.29})$$

$j_0$ is a saturation current density which represents the generation/recombination rate of holes in the bulk of the semiconductor as given by

# Theory

$$j_0 = \frac{eDn_i^2}{N_D L_p} \tag{II.30}$$

where $n_i$ is the intrinsic electron density equivalent to $(n_0 p_0)^{1/2}$; $N_D$ the density of donor states; the hole diffusion length $L_p$ is herein defined as $(D\tau)^{1/2}$. Meanwhile, the thickness of the space charge region, $d_{sc}$ can be estimated by

$$d_{sc} = \frac{1}{e}\left(\frac{2\Delta E_F \cdot \mathcal{E}_0}{N_D}\right)^{1/2} \tag{II.31}$$

To quantitatively sense the dimension of $d_{sc}$ using the dielectric constant $\varepsilon$=11.68 and a typical density of donor states $N_D$=10$^{17}$ cm$^{-3}$ of Si and assuming $\Delta E_F$=0.5 eV, one obtains that $d_{sc}$ is in the order of 0.1 µm. Within the space charge region where it is assumed that no recombination occurs, excitation of holes leads to the current

$$j_{sc} = eI_0[1 - \exp(-\alpha d_{sc})] \tag{II.32}$$

If recombinations at the semiconductor-solution interface are also neglected, the total hole current at the semiconductor surface is then given by the sum of $j_{\text{diff}}$ and $j_{sc}$. Together with Eq. (II.28), this summation leads to

$$\begin{aligned}j_v = j_{\text{diff}} + j_{sc} &= -j_0\left(\frac{p_d}{p_0} - 1\right) + \frac{eI_0 \alpha L_p}{1+\alpha L_p}\exp(-\alpha d_{sc}) + eI_0[1-\exp(-\alpha d_{sc})] \\ &= -j_0\left[\frac{p_s}{p_0}\exp\left(-\frac{e\Delta\phi_{sc}}{kT}\right) - 1\right] + \frac{eI_0 \alpha L_p}{1+\alpha L_p}\exp(-\alpha d_{sc}) + eI_0[1-\exp(-\alpha d_{sc})]\end{aligned} \tag{II.33}$$

Again, the subscript "v" in Eq. (II.33) indicates that the charge transfer occurs via the valence band. This current density must also be equal to the hole current derived from the surface hole densities. In a way similar to that used for deriving Eqs. (II.23) and (II.24), the anodic and cathodic current via the valence band can be obtained as

$$j_c^+ = ek_0 p_s C_{\text{red}} \exp\left[-\frac{(E_v^s - E_{F,\text{redox}}^0 - \lambda)^2}{4kT\lambda}\right] \tag{II.34}$$

$$j_c^- = ek_0 N_v C_{\text{ox}} \exp\left[-\frac{(E_v^s - E_{F,\text{redox}}^0 + \lambda)^2}{4kT\lambda}\right] \tag{II.35}$$

in which $p_s$ is the hole density at the surface of the semiconductor; $N_v$ is the density of states at the upper edge of the valence band; $E_v^s$ is the energy of the upper edge of the valence band at the surface of the semiconductor. At equilibrium in the dark, the anodic and cathodic currents are equal in magnitude, which can be regarded as a saturation current density $j_v^0$ determined by the rate of hole transfer at the interface via the valence band as follows:

$$j_v^0 = j_c^+\bigg|_{\text{equilibrium}} = ek_0 p_s^0 C_{\text{red}} \exp\left[-\frac{\left(E_v^s - E_{\text{F,redox}}^0 - \lambda\right)^2}{4kT\lambda}\right]$$

$$= j_v^-\bigg|_{\text{equilibrium}} = ek_0 N_v C_{\text{ox}} \exp\left[-\frac{\left(E_v^s - E_{\text{F,redox}}^0 + \lambda\right)^2}{4kT\lambda}\right] \quad \text{(II.36)}$$

Eq. (II.36) is subject to the condition $p_s = p_s^0$. The density of states at the upper edge of the valence band $N_v$ barely changes since the valence band is almost always full, i.e. $N_v$ is almost constant whether or not an equilibrium is achieved. In a non-equilibrium situation upon illumination, the net current density due to the hole transfer from the semiconductor surface to the solution can be obtained as

$$j_v = j_v^+ - j_v^- = j_c^+\bigg|_{\text{equilibrium}} \frac{p_s}{p_s^0} - j_v^-\bigg|_{\text{equilibrium}} = j_v^0\left(\frac{p_s}{p_s^0} - 1\right) \quad \text{(II.37)}$$

$p_s^0$ is related to the bulk hole density $p_0$ by the Boltzmann distribution function

$$p_s^0 = p_0 \exp\left(\frac{e\Delta\phi_{\text{sc}}^0}{kT}\right) \quad \text{(II.38)}$$

in which $\Delta\phi_{\text{sc}}^0$ is the potential drop across the space charge region under the equilibrium condition in the dark. From equations (II.37) and (II.38), one obtains

$$\frac{p_s}{p_0} = \left(\frac{j_v}{j_v^0} - 1\right)\exp\left(\frac{e\Delta\phi_{\text{sc}}^0}{kT}\right) \quad \text{(II.39)}$$

Inserting Eq. (II.39) into (II.33) leads to

$$j_v = \frac{j_0 + eI_0\left[1 - \frac{\exp(-\alpha d_{\text{sc}})}{1+\alpha L_p}\right] - j_0 \exp\left(-\frac{e\eta}{kT}\right)}{1 + \frac{j_0}{j_v}\exp\left(-\frac{e\eta}{kT}\right)} \quad \text{(II.40)}$$

17

in which $\eta = \Delta\phi_{sc} - \Delta\phi_{sc}^0$ is the overpotential created upon illumination; a new term $j_g$ called the generation current is defined as

$$j_g = j_0 + eI_0\left[1 - \frac{\exp(-\alpha d_{sc})}{1+\alpha L_p}\right] = j_0 + j_{ph} \qquad (II.41)$$

where $j_{ph}$ is the so called photocurrent density.

Equation (II.40), though rather complex in form, is of remarkable importance because it describes the overall charge transfer process via the valence band at an n-type semiconductor electrode for both anodic and cathodic polarizations. As mentioned earlier, $j_0$ represents the generation/recombination rate of holes in the bulk of the semiconductor and $j_v^0$ the rate of hole transfer at the interface. The ratio $j_0/j_v^0$ indicates whether the generation/recombination or the surface kinetics of the hole transfer is rate determining. If $j_0/j_v^0 \gg 1$, i.e., the rate is controlled by surface kinetics due to slow hole injection, then

$$j_v = j_v^0\left[\exp\left(-\frac{e\eta}{kT}\right) - 1\right] + j_{ph}\frac{j_v^0}{j_0}\exp\left(-\frac{e\eta}{kT}\right) \qquad (II.42)$$

which leads to a constant current $-j_v^0$ under high negative polarization in the dark. If $j_0/j_v^0 \ll 1$, i.e. the current is controlled by the generation/recombination rate, then

$$j_v = j_0\left[1 - \exp\left(-\frac{e\eta}{kT}\right)\right] + j_{ph} \qquad (II.43)$$

It is straightforward that under high positive polarization, the current levels can be expressed as the sum of $j_0$ and $j_{ph}$. Since the charge carriers to be transferred across the interface are minority carriers, $j_0$ is usually of very small magnitude and depends on material properties such as diffusion coefficient and the diffusion length of minority carriers, as expressed in Eq. (II.30). For instance, using typical values of the parameters in Eq. (II.30) for Si electrodes, the dark current density $j_0$ is given in the order of $10^{-12}$ A·cm$^{-2}$, which is a hardly detectable value. Such small currents can be considerably enhanced by light excitation. The limiting value is obtained once all the excited minority carriers reach the surface where they are consumed in the reaction process. However, this is still a simplified analysis of charge transfer at the illuminated semiconductor/electrolyte interface because, for example, the cathodic dark current is hitherto assumed to be only due to the injection of holes into the valence band of the n-type semiconductor. More complexities will arise, if other effects such as surface states are introduced, which will not be further discussed here.

## II.1.4. Photocatalytic reactions by charge transfer at semiconductor nanoparticles

During the last decades, many investigations have been performed with semiconductor particles, either dissolved as colloids or used as suspensions in aqueous solutions. In photocatalytic reactions, the essential advantage of using semiconductor particles is their large surface area because in principle, the same reactions should occur at particles and extended electrodes. Besides, the photogenerated charge carriers can easily reach the surface before they recombine, so that a high quantum yield can also be expected. However, two reactions, an oxidation and a reduction must proceed simultaneously at the same particle surface (otherwise the particle will be charged up, eventually leading to the stop of the overall reaction), as presented in Figure II-7. The slower process then determines the overall reaction rate. In this regard, the particle behaves practically as a microelectrode kept always under open circuit potential with the anodic and cathodic current being equal in magnitude. Using larger sized semiconductor particles, the partial currents in the dark are of rather limited magnitude under open circuit conditions, because the majority carrier (e.g. electrons for n-type semiconductor) density at the surface is small due to the depletion layer beneath the electrode surface, as indicated in Figure II-7A. In contrast, no space charge region is formed in much smaller particles of size $d \ll d_{sc}$ (Figure II-7B). Upon light excitation, some minority carriers (e.g. holes for n-type) in larger particles are transferred to the electron donor in the solution, which results in a negative charging of the particle that alleviates the positive space charge. Accordingly, this event causes a flattening of energy bands (see the dashed line in Figure II-7A), equivalent to a negative shift of the rest potential of a bulk electrode under illumination. Using much smaller semiconductor particles ($d \ll d_{sc}$), the photogenerated electrons and holes can be easily transferred to the surface and react with the electron and hole acceptors, as long as the energetic requirements are fulfilled. The average transit time $\tau_{tr}$ within a particle of diameter $d$ can be obtained by solving Fick's diffusion law as

$$\tau_{tr} = \frac{d^2}{4\pi^2 D} \tag{II.44}$$

# Theory

Taking typical values of $D$=0.1 cm²s⁻¹ and $d$=20 nm, the average transmit time is about 1 picosecond, which is much shorter than the recombination time so that most charge carriers can reach the surface before recombination.

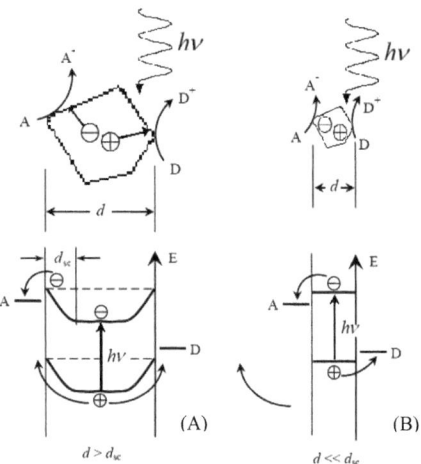

Figure II-7. Charge carrier transfer at large (A) and small (B) semiconductor particles in the presence of an electron donor D and an acceptor A [26].

Charge transfer reactions at extended electrodes have been mainly studied by using simple one step redox systems which are reversible. There is no point in studying corresponding reactions at particles, because a redox system being oxidized by a hole would be immediately reduced by an electron transfer from the conduction band. Therefore, only irreversible reactions of organic compounds have been investigated. Since small particles tend to conglomerate, stabilizers such as $SiO_2$ or polymers have been used. This may lead to problems in the case of a polymer, because it can be oxidized by holes. This problem can be avoided by using bare particles in solutions of a low ion concentration. Since the particles are charged, the counter charge extends relatively far into the electrolyte (Gouy Chapman layer), and conglomeration does not take place due to the repulsive forces between equally charged particles. When organic molecules, such as e.g. alcohols, are oxidized by hole transfer, usually oxygen in the solution acts as an electron acceptor. A whole sequence of reaction steps can occur which are frequently difficult to analyze, because also cross reactions may be possible and a new product is formed. One example is the formation of 2-phenylindazole from azobenzene at illuminated $TiO_2$-particles in methanol solutions. For instance, differences in the pathways of the photocatalytic reaction of acetic acid between a system using $TiO_2$ bulk electrode short circuited to a Pt electrode and that using Pt-loaded $TiO_2$ particles provide a very suitable illustration of such size dependent consequence [27]. Using spatially separated electrodes, acetic acid is oxidized by photogenerated holes at the illuminated $TiO_2$ electrode to form $CH_3^{\bullet}$ radicals which then combine with each other to yield ethane, while $H_2$ evolution is the result at the Pt electrode. Distinctively, at the Pt-loaded $TiO_2$ particles, a photogenerated electron (most likely trapped at Pt sites) and a hole are able to reduce a proton and oxidize acetic acid to yield surface-adsorbed hydrogen $H_{ad}$ and a $CH_3^{\bullet}$ radical, respectively. Then at adjacent $TiO_2$ and Pt sites, methane can be formed due to the reaction between neighboring $H_{ad}$ and a $CH_3^{\bullet}$ radical. Another interesting consequence of the size effect concerns the density of photons absorbed by semiconductor particles. Considering two colloidal solutions of different particle sizes and assuming that all photons are absorbed in both, the time interval between the absorption of two photons in the smaller particle solution can be exceedingly larger than that in the bigger particle solution. This difference can be very influential on these reaction pathways which require multiple electron transfers.

As mentioned above, the slowest reaction step determines the rate of the total reaction. For instance, if no electron acceptor is present in the solution, then the photoexcited electrons may be trapped in surface sites. Also the trapping of holes has been observed when electrons are efficiently scavenged. It is interesting to recognize that trapping of charges at the semiconductor surface upon illumination can be determined in particles by spectroscopic methods and at extended electrodes by capacity measurements. The position of energy bands at the surface of particles cannot be determined exactly, because capacity measurements are not possible. Their position can only be estimated by checking which reaction is possible. Frequently, methyl viologen ($MV^{2+}$) has been used as an electron acceptor which can accept an electron from the conduction band upon illumination, provided that the conduction band is above the reduction potential of $MV^{2+}$. The radical ($MV^{\bullet+}$) formed in this reaction is usually spectroscopically or electrochemically analyzed. These methods, however, give a very rough estimation, because usually it is not known whether surface states are involved in the charge transfer process. The band gap of semiconductor particles increases considerably when their size becomes smaller than about 100 Å. Accordingly, the position of energy bands is shifted, and it is expected that certain reactions should become possible with quantized particles which do not occur with bulk materials.

## II.1.5. Quantum size effect

As mentioned previously, reactions at extended electrodes and particles differ only insofar as, at particles, both an oxidation and a reduction process always occur simultaneously. Nanosized semiconductor particles have been very popular as photocatalysts due to their large surface area. Yet the most striking feature of the semiconductor nanoparticles is the remarkable change in their optical absorption spectra due to size reduction, compared with bulk materials [28]. For example, the band gap of CdS can be tuned between 2.5 and 4.5 eV as the size is varied from a macroscopic crystal down to the molecular regime. This interesting phenomenon has been addressed with success using the effective mass model. As mentioned earlier, in bulk semiconductors, light excitation results in the formation of electron-hole pairs, which experience a Coulomb interaction and can form excitons with a usually small bonding energy (<0.03 eV) and a large radius. The exciton radius is developed based on the Bohr radius of an electron in a H atom modified by introducing the semiconductor's dielectric constant and its reduced effective mass $m^*$ with the latter being given by

$$m^* = \left( \frac{1}{m_c^*} + \frac{1}{m_h^*} \right)^{-1} \tag{II.45}$$

Using a "particle in a box" model with an infinite potential drop at the wall as the boundary condition, and taking into account that the exciton consists of an electron-hole pair, the Schrödinger equation can be solved to yield the energy of the lowest excited state [29,30], i.e. the lower edge of the conduction band, as

$$E(R) \cong E_g + \frac{\hbar^2 \cdot \pi^2}{2 \cdot R^2} \left( \frac{1}{m_e^*} + \frac{1}{m_h^*} \right) - \frac{1.8 \cdot e^2}{4 \cdot \pi \cdot \varepsilon \cdot \varepsilon_0 \cdot R} + smaller\ terms \tag{II.46}$$

in which $m_0$ is the electron mass in vacuum. According to Eq. (II.46), when a semiconductor has a reduced effective mass which is significantly smaller than the free electron mass, large variation of its band gap can be expected. Examples of such semiconductors are given: CdS ( $m_e^*$ =0.21$m_0$, $m_h^*$ =0.80$m_0$), CdSe ( $m_e^*$ =0.13, $m_h^*$ =0.45 $m_0$), GaAs ( $m_e^*$ =0.067 $m_0$, $m_h^*$ =0.082 $m_0$) and ZnO

($m_e^*$ =0.24 $m_0$, $m_h^*$ =0.45 $m_0$) [31,32]. There are, however, semiconductors with larger effective masses due to which quantization does not occur theoretically. One example is $TiO_2$ which is crucial to realize in that $TiO_2$ has been the most intensively studied photocatalyst materials in the past decades. The effective electron mass in quantum sized $TiO_2$ particles has been reported to range between $5m_0$ and $13m_0$. Experimentally, $TiO_2$ particles synthesized with an average size between 5 and 20 nm were confirmed to exhibit the band gap properties of the bulk solid. However, when the $TiO_2$ particle size is controlled below 3 nm (i.e., corresponding to a few hundreds of $TiO_2$ molecules), quantum size effects could also be identified indicating a band gap increase of ~0.25 eV [33]. To precisely explore the quantum size effect, it should be emphasized that the band gap shifts can only be measured with sufficient precision employing colloidal suspensions, possessing a sufficiently narrow size distribution. Besides dispersed particles, techniques have also been available for fabrication of semiconductor films consisting of nanocrystalline particles [32]. Such films may exhibit similar quantum size effects as individual particles, depending on the effective mass of the semiconductor as just described. Accompanying the band gap widening due to reduced particle size, electrons at the lower edge of the conduction band and holes at the upper edge of the valence band then possess higher negative and positive potentials, respectively. In consequence, electrons and holes have a higher reduction and oxidation power, respectively in these quantized particles [26]. Besides band gap widening, variation of semiconductor particle size also has possible interesting consequences on the charge carrier transfer in photocatalytic reactions.

## II.2. Heterogeneous photocatalysis

Teichner and Formenti [34] described heterogeneous photocatalysis as an increase in the rate of a thermodynamically allowed ($\Delta G<0$) reaction in the presence of an irradiated solid with the increase (in rate) originating from the creation of some new reaction pathways involving photocreated species and a decrease of the activation energy. Semiconductors (e.g. $TiO_2$, $ZnO$, $Fe_2O_3$, $CdS$, and $ZnS$) can act as sensitizers for light reduced redox processes due to their electronic structure, which is characterized by a filled valence band and an empty conduction band [23]. When a photon with an energy of $h\nu$ matches or exceeds the band gap energy, Eg, of the semiconductor, an electron, $e_{cb}^-$, is promoted from the valence band, into the conduction band, leaving a hole, $h_{vb}^+$ behind. Excited state

# Theory

conduction band electrons and valence band holes can recombine and dissipate the input energy as heat, get trapped in metastable surface states, or react with electron donors and electron acceptors adsorbed on the semiconductor surface or within the surrounding electrical double layer of the charged particles.

In the absence of suitable electron and hole scavengers, the stored energy is dissipated within a few nanoseconds by recombination. If a suitable scavenger or surface defect state is available to trap the electron or hole, recombination is prevented and subsequent redox reactions may occur. The valence band holes are powerful oxidants (+1.0 to +3.5 eV vs. NHE depending on the semiconductor and pH), while the conduction band electrons are good reductants (+0.5 to -1.5 eV vs. NHE) [23]. Most organic photodegradation reactions utilize the oxidizing power of the holes either directly or indirectly; however, to prevent a buildup of charge one must also provide a reducible species to react with the electrons. In contrast, on bulk semiconductor electrodes only one specie, either the hole or electron, is available in the reaction due to band bending. However, in very small semiconductor particle suspensions both species are present on the surface. Therefore, careful considerations of both the oxidative and the reductive paths are required. The application of illuminated semiconductors for the remediation of contaminants has been used successfully for a wide variety of compounds such as alkanes, aliphatic alcohols, aliphatic carboxylic acids, alkenes, phenols, aromatic carboxylic acids, dyes, PCB's, simple aromatics, halogenated alkanes and alkenes, surfactants, and pesticides as well as for the reductive deposition of heavy metals (e.g. $Pt^{4+}$, $Au^{3+}$, $Rh^{3+}$ and $Cr^{6+}$) from aqueous solution to surfaces [23]. In many cases, complete mineralization of organic compounds has been reported. A general stoichiometry for the heterogeneously photocatalyzed oxidation of a generic chlorinated hydrocarbon to complete mineralization can be written as follows in equation (II.47):

$$C_xH_yCl + \left(x + \frac{y-z}{4}\right)O_2 \xrightarrow{h\nu, TiO_2} xCO_2 + zH^+ + zCl^- + \left(\frac{y-z}{2}\right)H_2O \qquad (II.47)$$

## II.2.1. Mechanisms

The experiments in the Friedmann et al., study involved irradiating a system containing a transparent $TiO_2$ colloid, silver ions, using silver perchlorate as precursor and polyvinyl alcohol (PVA) as the hole scavenger under varied photon fluxes with a laser flash photolysis set up [35]. These experiments showed the formation of metallic silver clusters on the surface of titanium dioxide immediately after irradiation with a laser pulse. These metallic silver clusters containing ≥ 12 silver atoms were observed using photon fluxes of $5\times10^{-5}$ and $1\times10^{-6}$ mol photon/L equivalent to approximately 60 and 2 photons per $TiO_2$ particle respectively. The corresponding quantum yields were calculated to be 0.3 and 0.5, respectively. Cluster growth was observed to occur in the millisecond range. The larger than expected silver clusters at the low photon flux, together with the high quantum yields, are believed to be a result of an antenna effect in which silver ions are reduced by electrons transferred through the self assembled $TiO_2$ aggregate. This requires that the silver metal deposits formed an ohmic contact with a highly doped $TiO_2$ semiconductor due to the creation of energy states as $Ag^+$ ions were chemisorbed and Ag nuclei were formed on the surface of $TiO_2$. Hence, this study reports the first observation of an antenna type effect for a photocatalytic reduction reaction.

In this postulation, the concept of the antenna effect which was introduced by Wang et al., [36] is extended to photocatalytic reduction reactions. The antenna effect involves the transfer of energy of a photon through a self assembled aggregate in which the particles have the same crystallographic orientation. The implications are that in a $TiO_2$ self assembled aggregate, a molecule adsorbed on a particle in this aggregate can still be oxidized, even when the particle on which it is located had not been directly activated by a photon (Figure II-8).

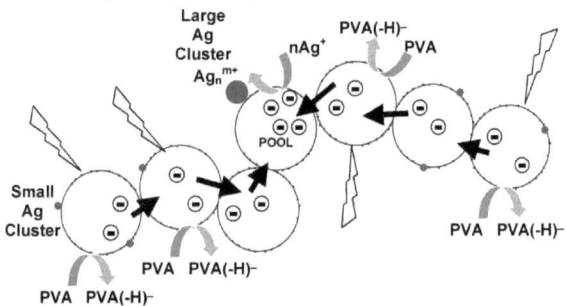

Figure II-8. Mechanism of reaction on the surface of $TiO_2$ with photodeposited Ag [35].

Ohno et al., observed the behaviour of an electron migration on rutile S-doped $TiO_2$-$Fe^{3+}$ particles under UV irradiation in presence of ethanol with a photoacoustic (PA) spectra [37]. The PA intensity indicates the amount of $Ti^{3+}$ ions generated by photooxidated electrons. Considering that the photoexcited electrons do not transfer to molecules such as $O_2$ molecules adsorbed on the surface of

TiO$_2$, the electrons are trapped in Ti$^{4+}$ sites in TiO$_2$ and reduce Ti$^{4+}$ ions to Ti$^{3+}$ ions in the bulk of TiO$_2$ or on the surface of TiO$_2$ (Figure II-9A).

The PA signal of S-doped TiO$_2$-Fe$^{3+}$ increased even under photoirradiation of about 470 nm and was much larger than that of S-doped TiO$_2$. These results suggest that metal ions absorbed visible light to generate excited states. The excited states of metal ions injected electrons into the S-TiO$_2$, resulting in reduction of Ti$^{4+}$ to Ti$^{3+}$. In addition, photoexcitation of metal ions adsorbed on the outside surface of S-doped TiO$_2$ occurred preferentially compared to that of S-doped TiO$_2$ under visible light irradiation (Figure II-9B). The mechanism of Ti$^{3+}$ generation can be explained in detail as follows: (1) photoexcited metal ions injected electrons into TiO$_2$ and the metal ions reached an oxidized state (M$^{(n+1)+}$); (2) the injected electrons reduced Ti$^{4+}$ to Ti$^{3+}$; and (3) the oxidized state of metal ions (M$^{(n+1)+}$) resulted in oxidation of ethanol to recover the initial state of metal ions (M$^{n+}$).

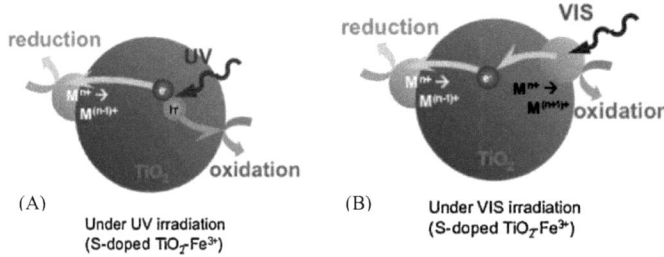

(A) Under UV irradiation (S-doped TiO$_2$-Fe$^{3+}$)

(B) Under VIS irradiation (S-doped TiO$_2$-Fe$^{3+}$)

Figure II-9. Mechanism of the reaction on the rutile S-doped TiO$_2$-Fe$^{3+}$ photocatalyst under UV (A) and visible (B) light irradiation [37].

## II.2.2. Stability problems

One requirement for the application of a semiconductor as a photocatalyst is the sufficient stability of the material or in other words the resistance to photocorrosion during the redox reactions. A semiconductor which doesn't present enough stability for example is CdS.

On the basis of the photoelectrochemical experiments with CdS single crystal electrodes can be understood that oxygen participates not only in the reduction but also in the oxidation process [38]. Reducing oxygen at clean CdS electrodes only leads to OH$^-$, and no peroxide could be found by

Meissner and Memming in the CdS<sup>-</sup> cathode compartment after electrolysis, but sulfate is found as the main photooxidation product of CdS in the presence of oxygen under anodic polarization:

$$O_2 + 4e^- + 2H^+ \longrightarrow 2OH^- \qquad (II.48)$$

photoprocess at illuminated CdS particles being given by

$$CdS + O_2 \xrightarrow{light} Cd^{2+} + SO_4^{2-} \qquad (II.49)$$

the well known anodic photocorrosion reaction of CdS in the absence of oxygen

$$CdS + 2h^+ \xrightarrow{light} Cd^{2+} + S \qquad (II.50)$$

Therefore, during the illumination of CdS suspensions in oxygen-containing solutions, hydrogen is evolved and oxygen is consumed. With increasing time of corrosion and with this increasing concentration of $Cd^{2+}$ the cathodic reaction

$$Cd^{2+} + 2e^- \longrightarrow Cd^0 \qquad (II.51)$$

and the anodic photocorrosion reaction in the presence of oxygen should finally lead to an overall stoichiometry of:

$$2CdS + 2H_2O + 3O_2 \xrightarrow{light} 2Cd^0 + 2SO_4^{2-} + 4H^+ \qquad (II.52)$$

As a mechanism for the photooxidation reaction of CdS in the presence of oxygen Henglein had assumed that peroxide may be the main oxidation product. Illuminating CdS colloids, he found small amounts of sulfate in the solution but at first attributed them at first to further oxidation of the corrosion product sulfur by $O_2^{\bullet-}$ or $H_2O_2$ in a following side reaction, though he modified this in later papers. Henglein took up Gerischer's idea of a radical ion being the first step in every photocorrosion process [39] and formulated as initial steps of the corrosion process. Meissner et al., presented from their photocarrosion experiments the overall reaction according surface-bound species from their photocorrosion experiments [38].

$$CdS + h^+ \longrightarrow Cd^{2+} + S^{\bullet-}_{(s)} \qquad (II.53)$$

$$S^{\bullet-}_{(s)} + O_2 \longrightarrow SO_2^{\bullet-}{}_{(s)} \qquad (II.54)$$

$$SO_2^{\bullet-}{}_{(s)} + h^+ \longrightarrow SO_{2\,(s)} \qquad (II.55)$$

$$SO_{2\,(s)} + H_2O + 2h^+ \longrightarrow SO_4^{2-}{}_{(s)} + 2H^+ \qquad (II.56)$$

$$\overline{CdS + 4h^+ + 2H_2O + O_2 \longrightarrow Cd^{2+} + SO_4^{2-} + 4H^+} \qquad (II.57)$$

## II.3. New materials for photocatalysis

There are many efforts done to modify the semiconductor structure of the nanomaterials with the intention of raising their absorption capacity to obtain a higher photoactivity. The crystallinity for example is one of the characteristics of the materials structure which can be modified by increasing the calcinations temperature. But there are other more complex modifications which are the photodeposition and the doping of materials which will be discussed later in detail. The main interest relies on the development of nanocompounds which should be able to absorb energy with lower wavelengths so that the solar energy could be used in an efficient way. The application of solar energy in common life would represent a very high impact in the economy of the world avoiding the production of several fuels, promoting less production of contaminants in the atmosphere and in the water. However one very important detail to be confirmed after the materials have been treated or modified is to make sure that the new material is stable and can resist the photocorrosion during the photocatalytic experiments.

### II.3.1. Photodeposited photocatalysts

The presence of co-catalytic noble metal deposits on metal-oxide particles is often employed to enhance the photocatalytic oxidation efficiency of processes through achieving a more efficient charge separation [40]. Considerable work has been done on Ag-$TiO_2$ modified photocatalysts and their potential applications as enhanced photocatalysts. The presence of Ag deposits on the $TiO_2$ surface can help to efficiently separate the electron-hole pairs by attracting the conduction band photoelectrons. This process has been shown to improve the overall efficiency for a number of photocatalytic reactions [41]. Ag-$TiO_2$ particles are typically prepared by photodeposition with the titanium dioxide particles being suspended in an aqueous solution containing the metal salt. When the suspension is illuminated, metal ions are reduced by conduction band electrons to form the metal deposits on the $TiO_2$ surface. This process is described by the series of reactions presented in II.58 to II.60. Depending on the preparation conditions, this photodeposition procedure typically yields small metal deposits ranging from a few nanometers to tens of nanometers [41].

# Theory

$$TiO_2 + h\nu \rightarrow TiO_2\,(e^-, h^+) \tag{II.58}$$

$$(TiO_2)OH + h^+ \rightarrow (TiO_2)OH^{\cdot\,+} \tag{II.59}$$

$$Ag^+ + e^- \rightarrow Ag^0 \tag{II.60}$$

The size and dispersion of metal deposits on $TiO_2$ particles are critical in controlling their photocatalytic activity. Given the strong effect of deposit size and dispersion, the development of preparation methods which provide such close control over deposit size is essential [42,43]. An issue that is not often addressed is the stability of these metal doped particles and hence the effects, if any, on the longer term application potential and shelf life stability of these particles. This is particularly true for applications as photocatalysts where these particles are subjected to irradiation with UV-A light. Reformation of surficial metal clusters as soon as the semiconductor particles are exposed to ultrabandgap irradiation has often been reported [44,45]. For example, while cluster stabilization can be achieved by adsorption of metal clusters onto a solid surface it has been shown that the rate of cluster growth on a quartz surface is accelerated by UV-A light [44].

The stability of metal deposits on photocatalysts under illumination may depend on a number of factors, such as, e.g. the type of metal, the deposit size, the photocatalytic reactions taking place, and the reaction conditions such as solution pH. Additionally, the redox potential can affect the reduction state of the supporting oxide. For example, the efficiency of the cluster oxide coupling has been found to strongly depend on the band bending in $TiO_2$ after silver deposition which in turn varies with the reduction state of the oxide [46].

Despite the large amount of studies focusing on metal deposited $TiO_2$ photocatalysts, particularly Ag-$TiO_2$, the photocatalyst stability and recycling has been addressed only in a few studies. Reproducibility tests were carried out on Ag-doped $TiO_2$ films prepared through a 2-step dipping and illumination process using Degussa P25 as the photocatalyst. These tests proved that the photocatalytic activity of the silver-modified films for the degradation of methyl orange remains unchanged even after six consecutive experiments with newly added pollutant quantities [47]. Xu and co-workers [48] have studied recycled 0.5 atom % Ag-deposited magnetic $TiO_2$-$SiO_2$-$Fe_3O_4$ photocatalysts and have also found that the photoreactivity for the degradation of orange (II) was maintained after 3 recycles. However, the total irradiation time was only 20 minutes which is relatively short.

In a study by Zhang, Ag-TiO$_2$ photocatalysts were in fact deactivated during the photocatalytic degradation of acetone in air [49], with a 73% reduction in activity during the fourth hour of reaction. The deactivation was explained as being due to interactions between partially oxidized organics and the Ag deposits, and possibly by the aggregation of the silver nanoparticles, which was reflected in changes in the absorption spectra of these particles. Regeneration was achieved by visible light illumination and was attributed to the oxidation of Ag deposits to Ag$^+$ ions, in the form of Ag$_2$O, which were reduced to Ag$^0$ upon UV illumination. This study raised the issue of the nanometer level control of feature size, inter-feature spacing and the long term stability of such structures, hindering their full exploitation for device applications [50].

Apart from photocatalytic applications, Sun et al. [51], have studied Ag-TiO$_2$ cyanide sensors prepared by the photodeposition method, which were tested under mild conditions and found to give good results using 0.1-10 µmol/L cyanide at pH 9-12. The Ag-TiO$_2$ sensors were found to be effective, without the need for regeneration for up to one month. It must be said that the available silver content was greater than the amount of cyanide to be detected.

In this study we have examined the durability of Ag-TiO$_2$ photocatalysts under practically relevant conditions. The model reaction was the photodegradation of dichloroacetic acid (DCA). DCA is a known industrial pollutant. Its photodegradation has been studied extensively, e.g., by Bahnemann et al. [52]. Important experimental parameters were selected for an initial exploration of the stability like the duration of illumination and the silver content. The poisoning of the photocatalyst surface by chloride ions was explored and reactivation was investigated using a simple washing technique. The deactivation of self-prepared colloidal TiO$_2$ was compared to that of the commercially available P25 TiO$_2$ photocatalysts.

## II.3.2. Doped material

Different wide band gap semiconductor materials have been applied as active photocatalysts. However, this wide band gap only allows very limited utilization of solar energy. Sensitization by dye molecules has shown encouraging results in dye-sensitized solar cells [53,54]. Photosensitization is defined officially, as a process whereby a photochemical change occurs in one molecular entity as a result of initial photon absorption by another molecular species. However in most photocatalytic systems the dyes are usually not stable once the photocatalyst is exposed to band gap illumination. Another way to shift the absorption towards the visible region is to dope the semiconductors with ionic species.

Often it is an advantage to be able to manipulate the equilibrium optoelectronic properties of a semiconductor. This can be achieved by deliberate introduction of impurities of a semiconductor. This impurity can occupy either a lattice site, an interstitial site, or it can simply be a vacancy. In most applications, a foreign element, known as a dopant, is added to a semiconductor to change the equilibrium concentration of electrons or holes.

A semiconductor that has been doped is known as an extrinsic semiconductor. If the semiconductor is doped with donors, it is referred to as an n-type semiconductor, and if the solid is doped with acceptors it is called a p-type semiconductor.

Donor and acceptor energy levels are known for a wide variety of impurity semiconductor combinations. Most of the common dopant atoms have shallow energy levels and well defined effects. In some special cases, such as Si in GaAs, dopants are amphoteric, in that they can either be donors or acceptors, depending on the growth conditions.

The substitution of sulfur at oxygen sites can significantly modify the electronic structures of $TiO_2$ because a sulfur atom has a larger ionic radius compared to N and F. It was considered that the conventional techniques, such as the sol-gel method, the gas-phase treatment at high temperatures and physical or chemical deposition methods are difficult to dope sulfur into $TiO_2$ lattice due to the large formation energy required for the substitution of sulfur for oxygen [9].

The first reports about sulfur-doped $TiO_2$ were published by Umebayashi *et al.*, [55]. They prepared photocatalyst by calcination of $TiS_2$ at 500°C or 600°C under air. S-doped $TiO_2$ in anatase form was obtained. The residual sulfur atoms occupied oxygen sites in the $TiO_2$ lattice to form Ti-S bonds detected by XPS. Authors correlated sulfur doping with the shift of absorption edge of $TiO_2$ to a lower energy region. The colors of the samples were beige or white for samples prepared at 500°C and 600°C respectively. The sample, prepared by the calcination at 500°C for 90 min, had good photocatalytic properties in the visible light, in relation to MB degradation.

Umebayashi et al., also prepared S-$TiO_2$ by ion implantation and subsequent thermal annealing at 600°C in air [55]. In this rutile form of titania the sulfur atoms occupied oxygen sites to form Ti-S bondings, similarly as in the previous investigations. Authors observed the shift of absorption edge but did not examined the photoactivity of obtained samples. They observed the photon to carrier conversion during irradiation by visible light above 420 nm. S-doping into $TiO_2$ causes an increase in the width of the valence band, thus resulting in band gap narrowing.

Ohno et al., synthesized S-$TiO_2$ by mixing titanium isopropoxide with thiourea in ethanol [16,56]. Ethanol was removed by evaporation under reduced pressure. The powder was calcinated at various temperatures and yellow powders were obtained. Photocatalyst calcinated at 500°C for 3 h showed the highest activity according to methylene blue and 2-propanol. The surface areas for samples prepared at 400 and 500°C were 75.8 and 17.7 $m^2/g$. By XPS analysis they observed the S 2p (3/2) peak, which they attributed to $S^{6+}$ ions. After washing treatment with HCl solution or $Ar^+$ etching this peak was lowered but still visible. No peaks responsible for the presence of nitrogen and carbon were reported.

## II.3.3. Heat treatment for structure modifications

Important modifications, usually irreversible, emerge by simple temperature treatments in the synthesis of semiconductors. In order to obtain a highly active photocatalyst, it is necessary to optimize a number of its properties. Powders should have a large surface area to adsorb substrates and high crystallinity to diminish the electron-hole recombination [57]. The increase of calcination temperature causes the decrease of surface area, but simultaneously crystallinity increases [58]. It was reported that other parameters also play an important role in photocatalytic activity. The decrease of particle size influences an increase of surface active sites on a semiconductor surface which causes an increase of the amount of charge carriers [59]. This increase is limited by the critical value of the particle size (5-10 nm). Below this value, the rate of electron-hole recombination

# Theory

increases. Another important property is the density of surface hydroxyl groups. Arslan et al., reported that the increase of surface area influences the increase of this density [60]. Oxygen defects localized on the semiconductor surface improve water adsorption and also increase the density of surface hydroxyl groups [10]. An acidity of $TiO_2$ surface can contribute the overall activity of prepared photocatalyst. For each preparation method, the optimal calcination temperature to achieve good adsorption properties and good crystallinity should exist.

It is well known that anatase is the more photoactive form of $TiO_2$ than rutile. Rutile is usually prepared in high temperatures (above 700 °C), therefore its surface area is low and the crystalline size is large as a result of agglomeration. Phase transformation from anatase to rutile causes the surface dehydroxylation and the $e^-/h^+$ recombination rate increases [61]. The photocatalyst Aerosil P25 from Evonic Degussa is a mixture of 80% anatase and 20% rutile, but it has been proven that there is no synergetic effect between these two $TiO_2$ phases. In this case, the relation of the surfaces and structures after the calcination of anatase in order to produce rutile is a good example of how a simple heat treatment can increase the photoactivity of a photocatalyst, by mixing two structures of the same photocatalyst. Another interesting semiconductor for example is the $Ga_2O_3$ which can be turned from an $\alpha$-$Ga_2O_3$ phase to a $\beta$-$Ga_2O_3$ after temperature treatments. These structures are quite different, as will be discussed later. These changes in the structure represent intrinsic modifications generating oxygen or gallium vacancies, depending on the treatment, which can also change the photocatalytic performance.

## II.4. Diverse photocatalytic test systems

There are different photocatalytic applications as it can be observed in Figure II-10. However, in this work particularly the applications related to the water treatment and air cleaning effect were studied. Even though, it is interesting to mention shortly the self-cleaning effect as another use of the $TiO_2$ photocatalyst in common life.

Typical photocatalytical test systems are divided by decomposing specific model compounds in each of the tree different phases: solid, liquid and gas. For the solid phase is developed the so called "self-cleaning effect", explained below (Figure II-11), is developed.

As a water treatment example the decomposition of dichloroacetic acid (DCA) was practiced, considering DCA as a model pollutant in aqueous phase. Other experiments for the gas phase treatment studies were followed separately by the photochemical decomposition of two different substances commonly present in contaminated air, an organic and an inorganic compound, acetaldehyde and $NO_x$ respectively.

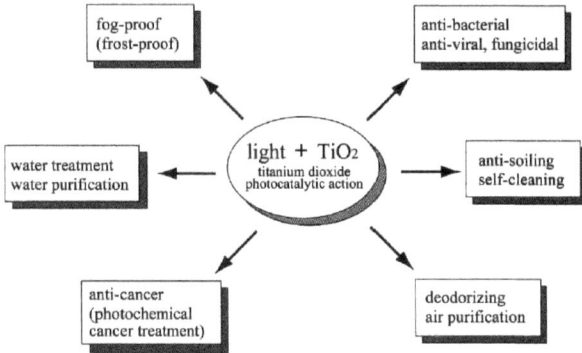

Figure II-10. Photocatalytic applications [62].

## II.4.1. Self cleaning effect

A self cleaning function can also be expected with $TiO_2$-coated substrates when small amounts of organic contaminants accumulate gradually on it. Although it is difficult to analyze this effect quantitatively, such a self-cleaning system is only anticipated under conditions, in which the number of molecules of staining substance, reaching the surface is much lower than the number of photons (Figure II-11). However, the $TiO_2$-coated substrate shows unexpectedly effective self-cleaning effects in both ordinary indoor living and outdoor spaces after irradiation [26].

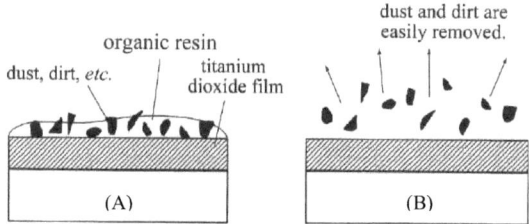

Figure II-11. Before irradiation (A) and after irradiation (B) [62].

# Theory

To explain one important effect, which results in the remotion of the adhesive forces between solid and liquid, the degree of water repellency has been considered (Figure II-12). It can be expressed in terms of contact angle of a water drop with the surface. On glass or other inorganic materials, water has a contact angle ranging from 20° to 30°. With plastics, the contact angle is typically from 70° to 90° (Figure II-13A). Very few substances show angles lower than 10°, only those that have been activated for example by soap. But these surfaces do not retain this effect for longe terms. Interestingly, a photogenerated amphyphylic effect has been discovered. This is an UV light irradiated surface of $TiO_2$ which generates a hydrophilic and oleophilic phase (Figure II-13B). Due to the effect of superhydrophilicity a continuous water film layer instead of droplets is formed.

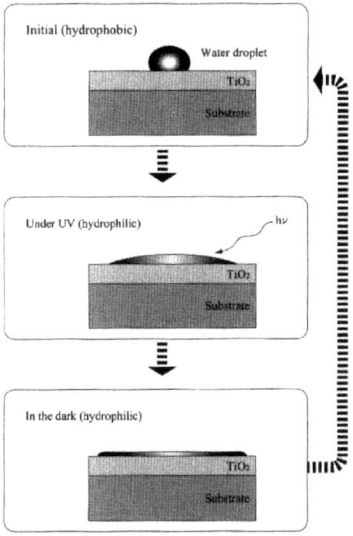

Figure II-12. Superhydrophilicity occurs under light irradiation [62].

Theory

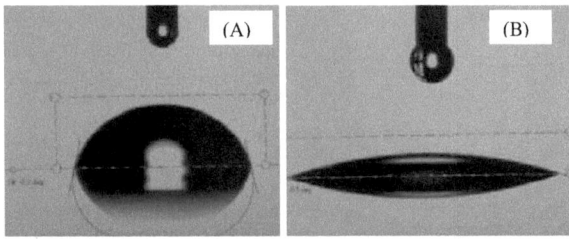

Figure II-13. Hydrophobic (A) and after light irradiation hydrophilic and oleophilic (B) (Photo, Königs).

## II.4.2. Photocatalytic decomposition of DCA on bare TiO$_2$ in aqueous phase

The model compound chosen for the photoactivity tests performed was dichloroacetic acid (DCA), which has also been studied before [63,64]. DCA has specific properties as a pollutant, which are useful for its decomposition analysis. It is a real toxic and carcinogenic molecule which doesn´t absorb light, is a stable compound difficult to be decomposed but has the capacity of being completely dissolved in water and the content of chloride ions in DCA can also simulate chlorinated water by its decomposition. This result is relevant, because chlorination of waters is the most used actual water treatment all over the world. However, a high value of Cl$^-$ ions concentration in water is confirmed to be carcinogenic. Therefore the use of photocatalysis TiO$_2$ appears to be a convenient pretreatment for potable water since high concentrations of Cl$^-$ ions can be attached to the photocatalyst surface and removed. The interesting part as presented in this work, is the photoactivity of the photocatalyst which can be recovered after washing the TiO$_2$ particles in order to be reused. The decomposition of DCA can easily be detected by three independent analytical methods (explained in detail below) to stoichiometrically confirm each of the results of the degradation products presented in equation (II.61):

$$CHCl_2COO^- + O_2 \xrightarrow{hv, TiO_2} 2CO_2 + H^+ + 2Cl^- \qquad (II.61)$$

The following schema in Figure II-14 describes the mechanism reaction of the degradation of the model compound DCA. Where the UV wavelength of the light reaches the TiO$_2$ particle surface the

# Theory

particle polarizes with a positive side and a negative one, so that the oxidation and reduction reactions can take place respectively.

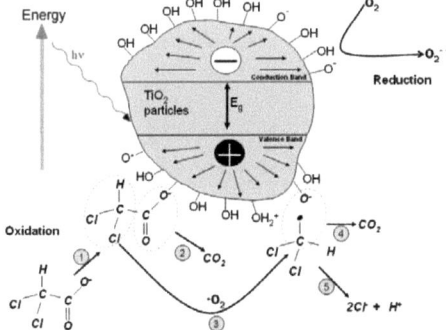

Figure II-14. Mechanism of DCA degradation of the photocatalytic reactor set up.

A brief explanation concerning the three experimental analyses is presented in Figure II-15. In the first analysis, one mMol decomposed of DCA resulted in one mMol of released protons [H$^+$] and was measured with a pH electrode. The pH electrode was connected to and regulated by a pH-stat-system to maintain a constant pH and also was connected to a computer to record the total [H$^+$] produced by the reaction. In the second analysis, two mMols of chloride ions [Cl$^-$] were measured continuously with a selective chloride ion electrode immersed in the photocatalyst suspension slurry. Finally, a third measurement confirms the results. It consisted in taking samples during the reaction to measure the total organic carbon (TOC) concentration [ppm] remaining in the slurry until the end of the reaction. These tests compared to each other confirmed the improvement of the photocatalysts under UV and/or visible light irradiation.

# Theory

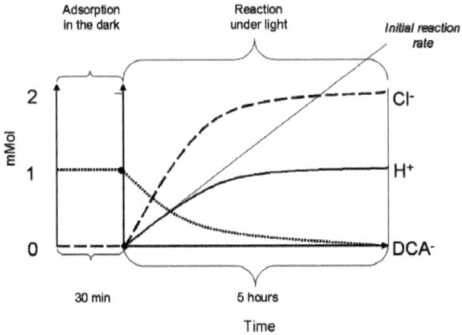

Figure II-15. Schematic diagram of the DCA degradation.

## II.4.3. Photocatalytic gas phase decomposition

In the gas phase two different gases were studied separately. $NO_x$ and acetaldehyde which were decomposed in flow reactor systems under UV-A or visible light depending on the photocatalyst tested.

Nitrogen oxides ($NO_X$, NO and $NO_2$) are hazardous air pollutants which are emitted from automobiles and combustion facilities. High $NO_x$ concentrations are often observed along highways and reduction of the $NO_x$ level is strongly requested in large cities (Figure II-16). The photocatalytic ability of thin films is almost the same as that of a commercial photocatalyst (P25; Nippon Aerosil) [65], at the same weight (about 60 mg) and illuminated area (100 cm²) of thin films. NO is converted to $HNO_3$ with photooxidation by way of $NO_2$.

$$NO + \frac{1}{2}O_2 \xrightarrow{light} NO_2 \qquad (II.62)$$

$$2NO_2 + \frac{1}{2}O_2 + H_2O \xrightarrow{light} 2HNO_3 \qquad (II.63)$$

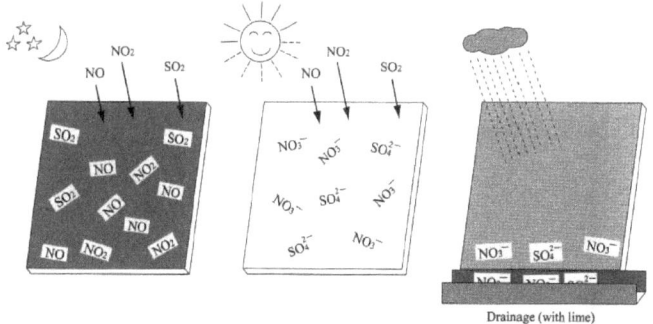

Figure II-16. $NO_x$ and $SO_x$ can be removed from the environment through photocatalysis [62].

The decrease in photocatalytic activity of the catalysts is attributed to the coverage of the catalyst surface by the $NO_2$ or $HNO_3$ formed [65].

The decomposition of acetaldehyde is more complex due to the produced intermediates. After the photogenerated $^\bullet OH$ radical from the $TiO_2$ surface it reacts with the acetaldehyde abstracting a hydrogen atom to form $CH_3C^\bullet O$ [66]:

$$CH_3CHO + {}^\bullet OH \rightarrow CH_3C^\bullet O + H_2O \qquad (\text{II}.64)$$

The oxygen present will attack the radical formed, producing an unstable peroxo radical.

$$CH_3C^\bullet O + O_2 \rightarrow CH_3(CO)OO^\bullet \qquad (\text{II}.65)$$

The $CH_3(CO)OO^\bullet$ radical reacts with another acetaldehyde molecule subsequently generating acetic acid and the $CH_3C^\bullet O$ radical,

$$CH_3(CO)OO^\bullet + CH_3CHO \rightarrow CH_3(CO)OOH + CH_3C^\bullet O \qquad (\text{II}.66)$$

$$CH_3(CO)OOH + CH_3CHO \rightarrow 2CH_3(CO)OH \qquad (\text{II}.67)$$

# Theory

This reaction constitutes a chain reaction mechanism so that CH₃C•O is recycled back to the equation (12) and the $CO_2$ is generated from acetic acid as follows:

$$3CH_3CHO + O_2 + {}^\bullet OH \rightarrow 2CH_3(CO)OH + CH_3C^\bullet O + H_2O \qquad \text{(II.68)}$$

$$CH_3(CO)OH + {}^\bullet OH \rightarrow CO_2 + H_2O + CH_3 \qquad \text{(II.69)}$$

# III. Materials and Methods

Devices, compounds and materials for the preparation or analysis of this investigation are presented in this part.

### III.1. For deposition synthesis of Ag on $TiO_2$

The photochemical experiments were realized in a reactor with a plain quartz window on which the light beam ($\lambda \geq 320$ nm) was focused. The photoreactor was filled with 50 mL of the aqueous suspension containing 0.5 g/L $TiO_2$ photocatalyst at different silver loadings from 0 to 2 atom%. The photocatalyst slurry was preilluminated with UV-A light for 30 minutes to oxidize any possible organic carbon impurities on the Ag-photodeposited $TiO_2$ surface before the decomposition reaction could take place by the addition of 1 mMol DCA. Silver perchlorate hydrate ($AgClO_4$) 99% purity from Aldrich was used as silver precursor. The preparation process for the deposition of silver on the $TiO_2$ surface allowed 24 hrs of mixing under 1 mW UV-A light illumination of the previously defined and added silver atom% particles on the suspending 500 mg P25 $TiO_2$ photocatalyst. It was in the presence of 20 μL methanol within a total volume of 1L of a 10 mMol $KNO_3$ water dissolution.

The prepared particles were characterized by electron microscopy (Emission Transmission Electron Microscope JEOL JEM-2100F-UHR and Emission Scanning Electron Microscope JEOL JSM-6700F). The STEM was supported by an energy dispersive spectroscopic analysis (EDX) and elemental x-ray-mapping images provided information, regarding the distribution of the various components ($TiO_2$ and Ag) and ions (Ti, Ag, Cl) before and after the photoreaction.

UV-Vis absorbance measurements in the wavelength range 200-700 nm showed the steep increase in absorption below approximately 350 nm, which is typical for nanosized $TiO_2$.

### III.2. For the S-doped $TiO_2$-$Fe^{+3}$ photocatalyst

Titanium dioxide ($TiO_2$) powder having an anatase phase was obtained from Ishihara Sangyo (ST-01). The relative surface area of ST-01 was 285.3 $m^2/g$. 2-Propanol, $FeCl_3$ and acetone were obtained from Wako Pure Chemical Industry. Thiourea and urea were obtained from Tokyo Chemical Industry Co., Ltd. Other chemicals were obtained from commercial sources as guaranteed reagents and were

## Materials and Instruments

used without further purification. The crystal structures of $TiO_2$ powders were determined from X-ray diffraction (XRD) patterns measured by using an X-ray diffractometer (Philips, X'Pert-MRD) with a Cu target K$\alpha$-ray ($\lambda$ = 1.5405 Å ). The relative surface areas of the powders were determined by using a surface area analyzer (Micromeritics, FlowSorb II 2300). The absorption and diffuse reflection spectra were measured using a Shimadzu UV-2500PC spectrophotometer. X-ray photoelectron spectra (XPS) of the $TiO_2$ powders were measured using a JEOL JPS90SX photoelectron spectrometer with an Al-K$\alpha$ source (1486.6 eV). This powder was called S-doped $TiO_2$, because a small amount of S atoms, but no C and N atoms are included in it.

### III.3. For the synthesis of indium selenide

For the synthesis of $In_xSe_y$ indium acetate (99.99 %), hexadecylamine (90 %), dodecylamine (98 %), sulfur (99.5 %), selenium (99.5 %), methanol (99.9 %), chloroform and acetone were used as received and were obtained from (Aldrich). Powder XRD data were collected using a Bruker D8 AXE diffractometer (Cu-K$\alpha$). The prepared particles were characterized by electron microscopy (Emission Transmission Electron Microscope JEOL JEM-2100F-UHR and Emission Scanning Electron Microscope JEOL JSM-6700F). The STEM was supported by an EDX analysis and elemental x-ray-mapping images provided information regarding the distribution of the components (In and Se).

### III.4. For the synthesis of beta gallium oxide

To produce $\beta$-$Ga_2O_3$ nanoparticles were required for the preparation of a self synthesized gallium acetate and compounds like hexadecylamine (90 %), dodecylamine (98 %), methanol (99.9 %), chloroform and acetone which were used as received were purchased from Aldrich. For the gallium acetate synthesis gallium trichloride anhydrous (99.99 %), glacial acetic acid and ether anhydrous were combined and mixed for a long period of time at constant high temperatures.
Powder XRD patterns were collected and EDXS analysis was done to prove the existence of beta gallium oxide.

## III.5. Set-up for photoactivity tests in aqueous phase, DCA degradation

The photochemical reactions were performed either in a 50 mL or in a 120 mL reactor with a quartz window on which the light beam of a Xe-lamp (OSRAM XBO-450W) was focused (Figure III-1A) and (Figure III-1B) respectively.

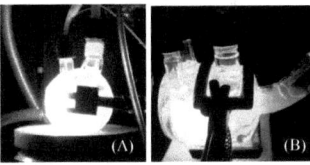

Figure III-1. Photocatalytic reactors (A) 50 mL and (B) 120 mL used for DCA degradation. Latter one designed to introduce a selective chloride electrode to follow Cl⁻ in-situ.

The Figure III-2 and the photo in Figure III-3 provide a schematic description of the photocatalytic reactor setup, which is explained in detail here, as the set-up for photoactivity tests in aqueous phase for DCA degradation. The infrared rays of the beam were absorbed by a water filter installed just in front of the reactor. The crossing UV-A visible light beam after the water filter was cut-off by two different filters. One filter with a wavelength $\lambda \geq 320$ nm was used for the UV light experiments and the second filter with a wavelength $\lambda > 420$ nm for the visible light experiments. The reactor was equipped with a magnetic stirring bar, a water-circulating jacket with openings for the pH electrode, chloride electrode, gas supply for bubbling the solution with air and a sample withdrawing. Depending on the photoreactor used, it was filled either with 50 mL or 120 mL of the aqueous suspension containing the photocatalyst.

Because the electron transfer is more efficient if the species are pre-adsorbed on a clean surface, the photocatalyst slurry was sonicated and subsequently pre-illuminated with near-UV light for 30 minutes to oxidize any organic carbon impurities on the $TiO_2$ surface before the photochemical reaction took place. Later on 1 mMol of DCA was added in the dark. The pH was then adjusted by addition of 0.1 M $HClO_4$ or 0.1 M NaOH, respectively. The temperature of the suspension was maintained at 25°C and was vigorously stirred without illumination for 30 minutes to attain adsorption equilibrium of the DCA molecules on the photocatalyst surface and continuously purged with air to ensure a constant $O_2$ concentration throughout the experiment. After this adsorption period, the reaction started by switching on the light. The intensity of UV-A illumination was 20 mW/cm² at the

## Materials and Instruments

entrance window of the photoreactor as measured by a UV light meter (Ultra-Violet Radiometer LT Lutron UVA-365).

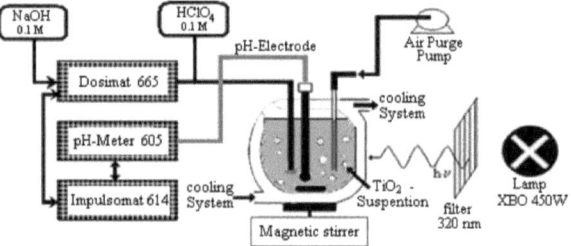

Figure III-2. Schematic of the photocatalytic reactor set-up.

Figure III-3. Picture of the photocatalytic pH-stat system.

The p$K_a$ of DCA is 1.29 [67] and it thus exists in its anionic form in aqueous solutions at pH > 2. The photocatalytic oxidation of one DCA anion results in the formation of one proton, two $CO_2$ molecules and 2 $Cl^-$ ions (Eq. II.61). The reaction pH was maintained constant using a pH-stat setup with the addition of $HClO_4$ or NaOH as needed.

The rate of the photodegradation of DCA was followed by measuring the amount of $OH^-$ added to keep the pH constant and thus the amount of $H^+$ formed using the pH-stat technique as described by Bahnemann et al., 1993 [68]. In total, three consecutive runs were preformed continuously with the same photocatalyst in order to test its stability. 1 mM of DCA was injected at the beginning of each run.

## Materials and Instruments

### III.6. Measurements of pH, chloride ions and total organic carbon (TOC)

As mentioned erlier, the pH value of the different photocatalytic suspensions was followed by the release of $H^+$ in the several DCA decomposition reactions. The utilized pH electrode connected 8102 Orion Ross was connected to the pH-stat set up. The 120 mL reactor was designed in order to introduce a Sartorius PY-I04 electrode with which the concentration of the $Cl^-$ ions as a product of the decomposed DCA molecule could be followed. This electrode is a chloride combination ion selective electrode from Sartorius. The amount of organic carbon converted into carbon dioxide was measured employing a Shimadzu TOC-500 analyzer with NDIR optical system detector. The sample passes through a tube into the oven and is then calcinated at 680 °C, obtaining the total carbon by IR-measurements of carbon dioxide. The inorganic carbon, which can be found in the form of carbonate and hydrogen carbonate, first passed through a fresh phosphoric acid solution and then continued along initial route. Organic carbon is calculated as the difference between total carbon and inorganic carbon ($HCO_3^-$ and $CO_3^{2-}$ species). In samples taken at pH 3 the total organic carbon (TOC) values were directly measured then all carbon in the sample was converted to carbon dioxide and measured by infrared detection. The concentration of carbon dioxide was directly proportional to the peak intensity.

### III.7. Set-up for photoactivity tests in gas phase

The photocatalyst was preirradiated for 48 hours under an intensity of 10 $mW/cm^2$ with a UV-A lamp (Philips, Type HB541, Cleo Performance UV Type 3, 100 W) before it was used for any reaction. The purpose of the pre-illumination was to clean the photocatalyst surface of any adsorbed pollutant. The gas phase reactions with $NO_x$ or acetaldehyde were carried out in a hermetic acrylic reactor within a 39 $cm^2$ powder holder containing 4 g of pressed photocatalyst under a visible light source (halogen lamp type QT/DE-12, Shanghai Xiangshan-Lamp-Industrial Corp.) focused on the glass reactor window and with different filters in between. The $NO_x$ and acetaldehyde systems are in principle similar (Figure III-4), but with a specific analyzer detector in each case. The set up consists of a gas bottle containing either $NO_x$ (Figure III-5A) or acetaldehyde (Figure III-5B) which was regulated by a valve pressure. So that a mass flow controller (Brooks Instrument, Model 5850S) could establish 1 ppm as the initial pollutant concentration to be decomposed. The flow controller also set the 50 %

## Materials and Instruments

relative humidity. The dry air of the compressor and the wet air of the humidifier were mixed with the pollutant flow by a gas mixer. In the case of $NO_x$ the flow rate was 3 L/min and for acetaldehyde was 1 L/min. A bypass was installed in order to connect the gas mixer flow either to the reactor or directly to a gas chromatography analyzer (GC 955, Syntech Spectras) for measuring acetaldehyde or to a luminescence detector (Horiba, Model APNA-360) in the case of the $NO_x$ analysis. After the initial conditions were established and the steady state was reached in the system, the bypass was turned connecting the reactor to the analyzer. The gas flowed through the reactor with the pressed photocatalyst sample inside, without illumination for one hour to let the pollutant attain the adsorbed and desorbed equilibrium on the photocatalyst surface. By turning on the light the reaction initiated. The luminescence detector followed the concentration of $NO_x$, NO and $NO_2$ in the system. The data were recovered and displayed with a SMA-360 PC 1.2 program. In the case of the acetaldehyde system the data were stored in the GC analyzer.

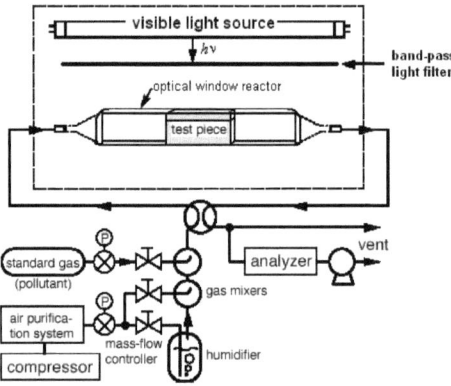

Figure III-4. Schematic presentation of the photocatalytic reactor set-up for gas phase.

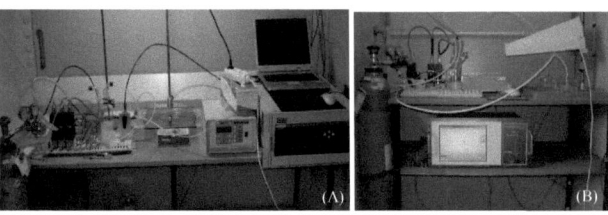

Figure III-5. Picture of the photocatalytic $NO_x$ (A) and acetaldehyde (B) gas phase set-up systems.

## III.8. Set-up for carrying out photocurrent measurements

The current-voltage characteristics of the deposited $In_2Se_3$ (0.25 cm$^2$ and 1 cm$^2$) or $Ga_2O_3$ (0.25 cm$^2$) electrodes, in contact with the sodium selenate $Na_2SeO_4$ (1 mM) electrolyte or with the $Na_2S_2O_3$ (0.1 M) electrolyte were measured in the dark as well as under illuminated conditions (80 mW/cm$^2$) by using a 320 nm cut-off filter (Figure III-6). The working electrodes and the platinum counter electrode were connected to a potentiostat Iviumstat Electrochemical Interface from Ivium Technologies applying a potential from -1.0 V to 1.0 V to compare the current and photocurrent potential values.

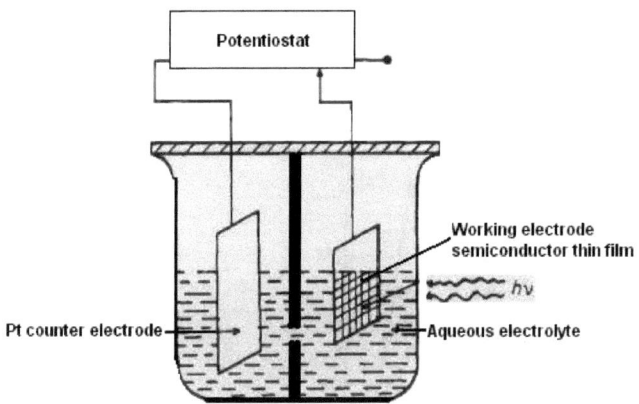

Figure III-6. Schematic presentation of the photocurrent set-up system.

## III.9. Set-up for cyclic voltammetry and Mott-Schottky measurements

Cyclic voltammograms were recorded with a computer connected to an E-Chemie $\mu$-Autolab potentiostat. This technique was used to study the electrode redox couple in a three electrode system conformed by the working electrode (semiconductor thin film of $In_2Se_3$ (0.25 cm$^2$ and 1 cm$^2$) or $Ga_2O_3$ (0.25 cm$^2$)), a platinum wire used as counter electrode and a saturated Ag/AgCl as reference electrode (Figure III-7). The experiments were performed in contact with the sodium selenate $Na_2SeO_4$ (1 mM) electrolyte or with $Na_2S_2O_3$ (0.1 M) electrolyte and were measured in the dark. The working electrode is swept between two sets of different potentials. From -2.0 V to 0.0 V for the

## Materials and Instruments

In$_2$Se$_3$ working electrodes and from -1.5 V to 2.0 V for the Ga$_2$O$_3$ working electrode. The scan rate (the magnitude of the change rate of the applied potential with time) is kept constant.

As the applied potential is increased, oxidation will occur and a positive current will flow (electron transfer from the species being oxidized to the electrode). Conversely, as the potential is decreased, reduction will occur and a negative current will flow (electron transfer from the electrode to the species being reduced). Reversible electrode dynamics are exhibited in systems where the heterogeneous rate constants are large [69].

The electrochemical set-up for the Mott-Schottky diagram construction consisted of three electrodes (Figure III-7): the working electrode (semiconductor thin film of In$_2$Se$_3$ or Ga$_2$O$_3$), a platinum wire used as counter electrode and a saturated Ag/AgCl electrode as reference connected to a potentiostat Iviumstat Electrochemical Interface from Ivium Technologies. The experiment was performed in an aqueous 0.1 M KCl solution at pH 7. The potential was systematically varied between -1.0 V and +1.0 V with the frequency range being modulated between 100 Hz to 1 kHz.

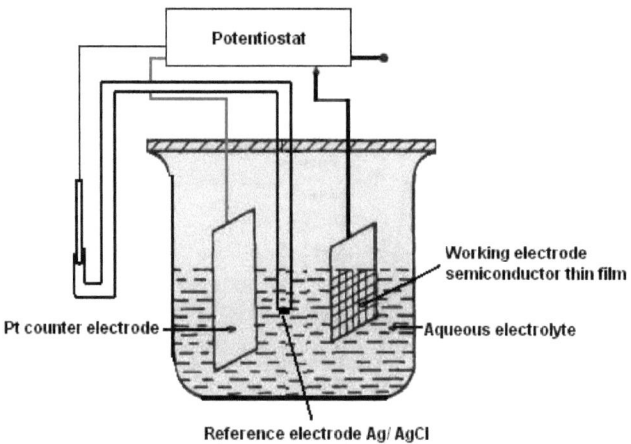

Figure III-7. Schematic presentation of the set-up system for voltammetry measurements and Mott-Schottky diagrams construction.

# IV. Results and Discussion

## IV.1. Photonic efficiency calculation

For a comparison between the performances of different photocatalysts by decomposing several pollutants in different phases, the calculation of the photonic efficiency has been applied considering different parameters as presented in this chapter.

### IV.1.1. Photonic efficiency calculation for UV-A light

The photonic efficiency $\xi$ is defined as the ratio of the initial degradation rate and the photon flux intensity (see equations IV.1 to IV.3), where the initial photocatalytic rate is calculated from the slope of the individual concentration versus the time profiles. This definition of the photonic efficiency was first suggested by Serpone et al. [4]. For example in the case of Ag-TiO$_2$ experiments the incident light intensity $I_o$ varied between 20 and 22 J*s$^{-1}$cm$^{-2}$. It was obtained based upon the UV-A light meter measurements [70]. The irradiated surface area was 8.042 cm$^2$ and the volume of the suspension was $5\times10^{-2}$ L.

$$\xi = \frac{Rate_{DCA}}{I} \cdot 100\ (\%) \qquad (IV.1)$$

$$Rate_{DCA} = k\left(\frac{1}{s}\right) \cdot C_O\left(\frac{mol}{L}\right) \qquad (IV.2)$$

$$I = I_O\left(\frac{J}{s \cdot cm^2}\right) \cdot \underbrace{\frac{1}{h \cdot c}\left(\frac{1}{J}\right)}_{\lambda} \cdot A\ (cm^2) \cdot \frac{1}{N_A}\left(\frac{1}{\frac{1}{mol}}\right) \cdot \frac{1}{V}\left(\frac{1}{L}\right) \qquad (IV.3)$$

With:
$I$ = photon flux intensity [mol* s$^{-1}$cm$^{-2}$]; $I_o$ = incident light intensity [J*s$^{-1}$cm$^{-2}$]; $\lambda$ = wavelength [nm]; $N_A$ = Avogadro's number [6.22*10$^{23}$mol$^{-1}$]; h = Planck constant [6.63*10$^{-34}$J*s]; c = light velocity [m*s$^{-1}$]; k = rate constant [s$^{-1}$]; A = area [cm$^2$]; $C_o$ = initial$_{DCA}$ concentration [mol*L$^{-1}$]; V = volume [L]

### IV.1.2. Photonic efficiency calculation modified for visible light intensity

The intention of doping the TiO$_2$ photocatalyst is in order to apply the visible light energy. However, after doping it and solving this first step related to the synthesis, the necessity of modifying the way to

## Results and Discussion

calculate the photonic efficiency was faced. Since it was not possible to directly measure the real intensity value of the visible light in the mW/cm² units, due to the lower energy, it was done with the UV-A lightmeter for the UV-A light. The photonic efficiency calculation for the visible light considered the very low amount of UV-A light energy measured with the UV-A lightmeter, but the highest energy contribution corresponds to the visible light intensity. The evaluation of the visible light is explained as follows. The visible light range represented by the area (53.545 a.u.) of the xenon lamp spectrum, where the doped material absorbs from 420 nm up to 555 nm (Figure IV-1), was multiplied by the relation (1.32*10⁻⁷ Einstein/s)/(29.273 a.u.) (Eq. IV.4) of the visible lamp spectrum without any filter (Table IV-1. & Figure IV-2). This result will be the energy related to the area of the xenon lamp spectrum for the same visible wavelength range. For example by comparing commercial photocatalysts to the S-TiO$_2$-Fe$^{3+}$ photocatalyst photonic efficiency results at the same experimental conditions, by the degradation of 1 mMol DCA using a cut off filter of 420 nm (Table IV-10) in order to activate the photocatalysts with visible light, the UV and visible light contribution were considered. In this case the photonic efficiency values of the sulfur doped photocatalyst obtained correspond to the 7.5*10⁻³ (mW/cm²) photons of the UV-A light measured by the UV-A lightmeter, which were able to come through the 420 nm cut off filter. However, this small quantity of UV-A photons has not the capacity for the found DCA degradation. This means that the reactions were photoactivated by real visible light photons, otherwise it is not possible to have a high conversion rate of reactants with such a low value of UV-A light.

$$(53.545\,a.u.) * \left( \frac{1.32 * 10^{-7} \frac{Einstein}{s}}{29.273\,a.u.} \right) = 2.41 * 10^{-7} \frac{Einstein}{s} \qquad (IV.4)$$

The proportion intensity factor between the xenon lamp and the halogen lamp was calculated considering the low UV-A emission measured from the xenon lamp 7.5*10⁻³ mW/cm² and the UV-A light photons 2.02*10⁻⁹ Einstein/s were calculated from the visible lamp (Eq. IV.5).

$$\frac{7.5 * 10^{-3} \frac{mW}{cm^2}}{2.02 * 10^{-9} \frac{Einstein}{s}} = 3.71 * 10^{6} \frac{\frac{mW}{cm^2}}{\frac{Einstein}{s}} \qquad (IV.5)$$

# Results and Discussion

To obtain the visible light energy of the xenon lamp, it is necessary to multiply the photon flux by the intensity factor of the visible lamp (Eq. IV.6):

$$\left(2.41*10^{-7} \frac{Einstein}{s}\right) * \left(3.71*10^{6} \frac{\frac{mW}{cm^2}}{\frac{Einstein}{s}}\right) = 8.96*10^{-1} \frac{mW}{cm^2} \qquad (IV.6)$$

The total incident photon flux, calculated for the xenon lamp was $2.127*10^{-7}$ mol·L$^{-1}$s$^{-1}$ which was obtained from the addition of the UV-A light ($7.5*10^{-3}$ mW/cm²) and the visible light ($8.96*10^{-1}$ mW/cm²) intensities at 420 nm.

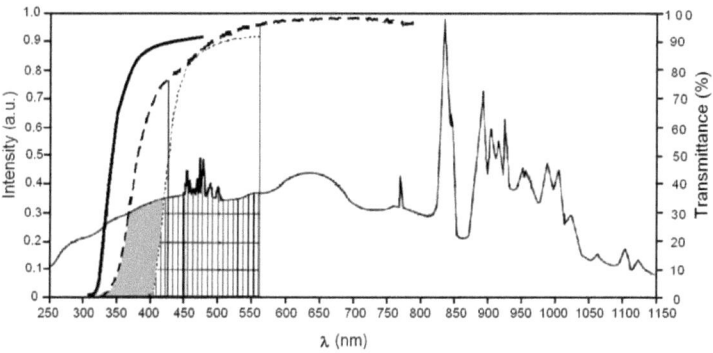

Figure IV-1. Relation between the S-doped $TiO_2$-$Fe^{3+}$ photocatalyst (---) and a xenon lamp (---)spectra to obtain the common area (53.545 a.u.; ▦) of the available visible light photons from 420 nm (cut off filter(···)) to the absorbing material limit 555 nm. For the UV light experiments the cut off filter 320 nm (—) was used.

## IV.1.3. Photonic efficiency calculation for the gas phase system

For the photonic efficiency calculation the S-doped $TiO_2$-$Fe^{3+}$ diffuse reflectance spectra and the lamp spectra were overlapped, to find out the shared area between both and it was applied a relation value for the available photonic energy (Figure IV-2). The light source was also measured under the different filters to detect spectra which could be compared to the photonic efficiency in-between. The used filters correspond to diverse commercial glass types with the purpose of proposing possible

## Results and Discussion

applications of the photocatalyst in the future. The glass slices which were used as filters were pieces of a common glass window, polycarbonate and a piece of a green Pilkington glass.

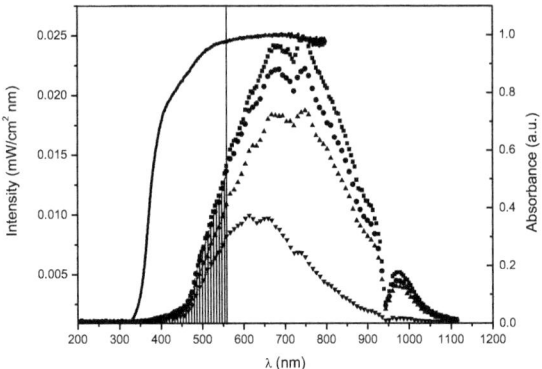

Figure IV-2. The shared area between the S-doped $TiO_2$-$Fe^{3+}$ diffuse reflectance spectrum (—) and the visible light source spectra with or without any filter from 400 nm to 555 nm (▥) (29.273) was obtain as a factor for further calculations of available visible light photons. The intensity without any filter (■) or with a glass filter (●) was 0.793 mW/cm$^2$; under a polycarbonate filter (▲) 0.642 mW/cm$^2$ and under a green filter (▼) 0.509 mW/cm$^2$.

The light intensities in Table IV-1 were calculated from the lamp emission spectra in Figure IV-2 according to the Annex 1 spectrum. The values correspond to the integration of the area under the respective spectra of each glass between the ranges of 360 nm to 400 nm for the UV-A light and from 400 nm to 555 nm for the visible light energy. The photon energy calculations were done in accordance to the sum of every 10 nm intensity value related to the area of interest.

Table IV-1. Visible light intensity values under the different filters for the gas-phase set up lamp.

| Filter used | UV-A light ($\lambda_{360-400\,nm}$) | | Visible light ($\lambda_{400-555\,nm}$) | | Total incident photon flux | |
|---|---|---|---|---|---|---|
| | (mW/cm$^2$) | (mol·L$^{-1}$s$^{-1}$) | (mW/cm$^2$) | (mol·L$^{-1}$s$^{-1}$) | (mW/cm$^2$) | (mol·L$^{-1}$s$^{-1}$) |
| Glass Window Reactor | 1.6*10$^{-2}$ | 2.02*10$^{-9}$ | 7.77*10$^{-1}$ | 1.32*10$^{-7}$ | 7.93*10$^{-1}$ | 1.34*10$^{-7}$ |
| Polycarbonate + Glass Window | - | - | 6.42*10$^{-1}$ | 1.09*10$^{-7}$ | 6.42*10$^{-1}$ | 1.09*10$^{-7}$ |
| Polycarbonate + Pilkington green glass + Glass Window | - | - | 5.09*10$^{-1}$ | 8.62*10$^{-8}$ | 5.09*10$^{-1}$ | 8.62*10$^{-8}$ |

## Results and Discussion

An example for the photonic efficiency calculation procedure of [NO] under visible-light using Pilkington glass green filter is described in this set of equations (IV.7 to IV.11).

$$\xi = \frac{Rate_{NO}}{I} \cdot 100 \ (\%) \tag{IV.7}$$

$$Rate_{NO} = \hat{n} \tag{IV.8}$$

$$\hat{V} = \Delta C_{NO}(ppm) \cdot Vol_{flux\,NO}\left(\frac{L}{s}\right) \tag{IV.9}$$

$$Rate_{NO} = \hat{n} = \frac{P\hat{V}}{RT} = \frac{1(atm) \cdot \hat{V}\left(\frac{L}{s}\right)}{0.082\left(\frac{L \cdot atm}{mol \cdot K}\right) \cdot 298(K)} = \left(\frac{mol}{s}\right) \tag{IV.10}$$

$$I = I_0\left(\frac{J}{s \cdot cm^2}\right) \cdot \frac{1}{\sum \frac{h \cdot c}{\lambda_i}}\left(\frac{1}{J}\right) \cdot A \ (cm^2) \cdot \frac{1}{N_A}\left(\frac{1}{\frac{1}{mol}}\right) = \left(\frac{mol}{s}\right) \tag{IV.11}$$

### IV.2. Degussa P25 as a standard photocatalyst

For the decomposition of different compounds in photocatalysis, the commercial photocatalyst Aerosil P25 from the company Evonic Degussa has been used as a comparison standard parameter for the different developed photocatalysts. Here it was employed for decomposing DCA in aqueous phase and $NO_x$ or acetaldehyde in the gas phase.

### IV.2.1. Degradation of DCA with bare Degussa P25 and its effect after washing

In Figure IV-3, the degradation of 1 mM DCA within each run (detected by the release of $H^+$ ions during the degradation process) using pure Degussa P25 as the photocatalyst is shown. The reaction time was 4 hours for each run. Three runs were consecutively carried out. A third run also presented here, was from a second series of experiments. Where 4 mM $Cl^-$ were added before the illumination to evaluate a possible poisoning effect of the photocatalyst's active sites by chloride ions. It can clearly be seen that both, the reaction rate and reaction yield are reduced during each consecutive run with the most obvious inhibition being observed upon the addition of chloride ions.

## Results and Discussion

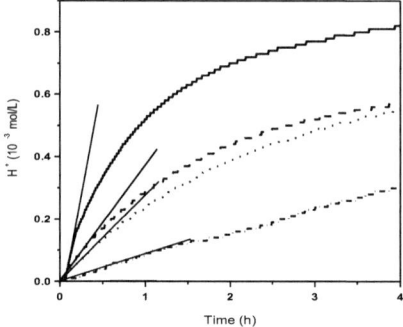

Figure IV-3. Degradation of DCA (shown as release of H⁺) using the photocatalyst Degussa P25 at pH 3 in 3 consecutive runs, 1st run (—), 2nd run (---), 3rd run (···) and a 3rd run (-··-) with addition of 4 mM Cl⁻ before the run started. The slope (—). used for the determination of the photonic efficiency of each run with $I \approx 3.39 \times 10^{-2}$ Einstein L⁻¹h⁻¹. The photocatalyst loading was 0.5 g/L.

As shown in Figure IV-4, the photocatalytic activity of Degussa P25 towards the degradation of DCA is almost completely recovered once the catalyst is simply washed with pure water, before the next activity. This clearly indicates the inhibitory effect of Cl⁻ ions, which was already obvious from the results presented in Figure IV-3.

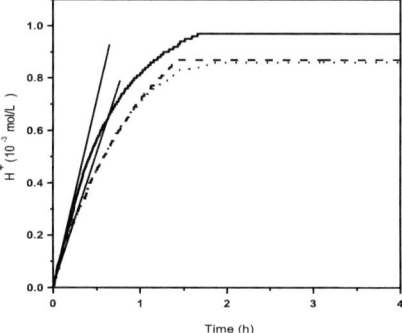

Figure IV-4. Degradation of DCA (shown as release of H⁺) using the photocatalyst Degussa P25 at pH 3 in 3 consecutive runs with intermittent washing between the runs, 1st run (—), 2nd run (---), 3rd run (···) and slope (—)

## Results and Discussion

used for the determination of the photonic efficiency of each run with $I \approx 3.39 \times 10^{-2}$ Einstein $L^{-1}h^{-1}$. The photocatalyst loading was 0.5 g/L.

### IV.2.2. Degradation of $NO_x$ with Degussa P25

Figure IV-5 shows the degradation of $NO_x$ experiments with Degussa P25 photocatalyst in gas phase under UV-A light. The progress of the $NO_x$ removal implies complex parallel reactions which are not analyzed in detail here. With a luminescence detector the $NO_x$ degradation can be followed as well as, the NO remotion and the produced $NO_2$ concentrations. The $NO_x$ and NO decompositions are represented by the values that are considered for the photonic efficiency degradation. In the graph the produced $NO_2$ is plotted to confirm the degradation of the initial $NO_x$ compound, however, since these are directly related to each other, it is not necessary to present further photonic efficiency analysis concerning $NO_2$.

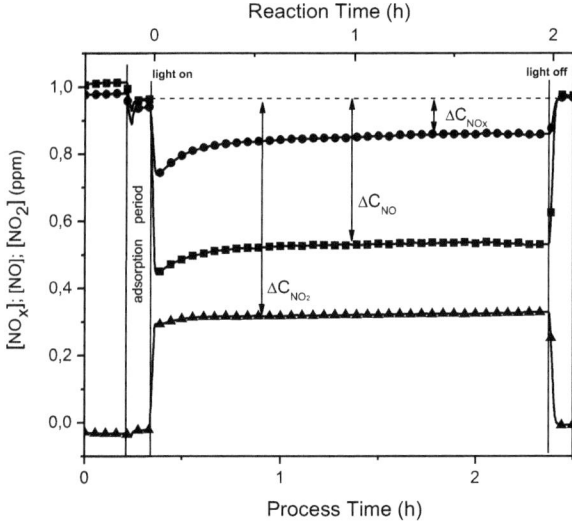

Figure IV-5. Comparison of 1 ppm NOx decomposition with 4 g P25 pressed powder photocatalyst applying 1 mW/cm² UV-A light without any filter. The reaction was followed by measuring $NO_x$ (●); NO (■); and $NO_2$ (▲).

## Results and Discussion

In Table IV-2 the results of the photonic efficiency, concerning the decomposition of $NO_x$ and NO with P25 as photocatalyst under UV-A light irradiation are presented.

Table IV-2. $NO_x$ and NO degradation results under UV light with P25 photocatalyst.

| Photocatalyst | (UV-light) | | | |
|---|---|---|---|---|
| | NO | | $NO_x$ | |
| | Rate [mol/s] | Intensity [mol/s] | Rate [mol/s] | Intensity [mol/s] |
| P25 (4.0g) | $8.859 \times 10^{-10}$ | $1.19 \times 10^{-7}$ | $2.189 \times 10^{-10}$ | $1.19 \times 10^{-7}$ |
| % Ph. Efficiency | 0.74 | | 0.18 | |

Furthermore, Degussa P25 as a standard photocatalyst was tested in order to compare the results with the ones of the S-doped $TiO_2$-$Fe^{3+}$ photocatalyst. The same experiment as the one presented here in Figure IV-6, was executed with S-doped $TiO_2$-$Fe^{3+}$ as photocatalyst (Figure IV-38) using the same conditions, however, different wavelengths were applied. A Pilkington green filter and a polycarbonate filter were used and they were compared to the same system without any filter. In this case by applying the light filters configuration the photonic efficiency results were much lower because the visible light photons could just attain to a very scarce NOx decomposition.

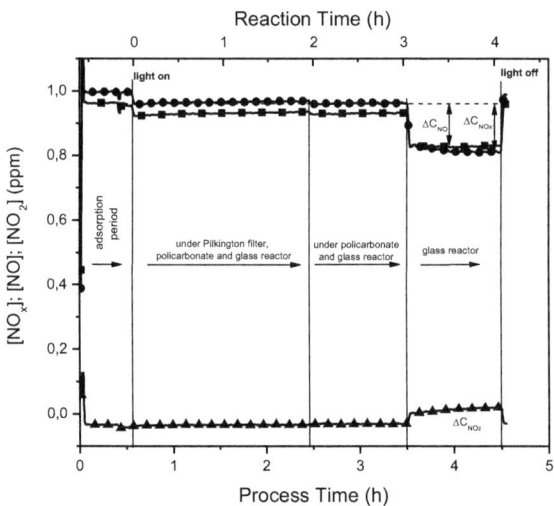

Results and Discussion

Figure IV-6. Comparison of 1ppm NOx decomposition with 4 g P25 pressed powder photocatalyst applying visible light photons under a Pilkington green filter, polycarbonate and without any filter. The reaction was followed by measuring NOx (●); NO (■); and $NO_2$ (▲).

The NO remotion results with Degussa P25 photocatalyst are presented in Table IV-3 and the respective results for the $NO_x$ degradation in Table IV-4. The degradation efficiency is lower under the Pilkington and polycarbonate filter configuration which means that P25 was able to absorb some of the visible photons reached but the values differ considerable to the ones obtained with the S-doped $TiO_2$-$Fe^{3+}$ photocatalyst.

Table IV-3. NO degradation results under visible light using different filters with P25 photocatalyst.

| Photocatalyst | $NO_x$ (VIS-light) | | | | | |
| --- | --- | --- | --- | --- | --- | --- |
| | Pilkington filter, Polycarbonate and glass reactor | | Polycarbonate and glass reactor | | glass reactor | |
| | Rate [mol/s] | Intensity [mol/s] | Rate [mol/s] | Intensity [mol/s] | Rate [mol/s] | Intensity [mol/s] |
| P25 (4.0g) | $0.652*10^{-10}$ | $8.62*10^{-8}$ | $0.767*10^{-10}$ | $1.09*10^{-7}$ | $3.9*10^{-10}$ | $1.34*10^{-7}$ |
| % Ph. Efficiency | 0.075 | | 0.07 | | 0.292 | |

Table IV-4. $NO_x$ degradation results under visible light using different filters with P25 photocatalyst.

| Photocatalyst | NO (VIS-light) | | | | | |
| --- | --- | --- | --- | --- | --- | --- |
| | Pilkington filter, Polycarbonate and glass reactor | | Polycarbonate and glass reactor | | glass reactor | |
| | Rate [mol/s] | Intensity [mol/s] | Rate [mol/s] | Intensity [mol/s] | Rate [mol/s] | Intensity [mol/s] |
| P25 (4.0g) | $0.517*10^{-10}$ | $8.62*10^{-8}$ | $0.636*10^{-10}$ | $1.09*10^{-7}$ | $2.62*10^{-10}$ | $1.34*10^{-7}$ |
| % Ph. Efficiency | 0.06 | | 0.058 | | 0.196 | |

## IV.2.3. Degradation of acetaldehyde with Degussa P25

Figure IV-7 shows the degradation of acetaldehyde with Degussa P25 photocatalyst in the gas phase under UV-A light. The progress of the decomposition of acetaldehyde was followed with a gas chromatograph detector. It can be observed. A high affinity between the photocatalyst surface and the acetaldehyde compound, due to the big adsorption peak during the dark period, can be observed.

## Results and Discussion

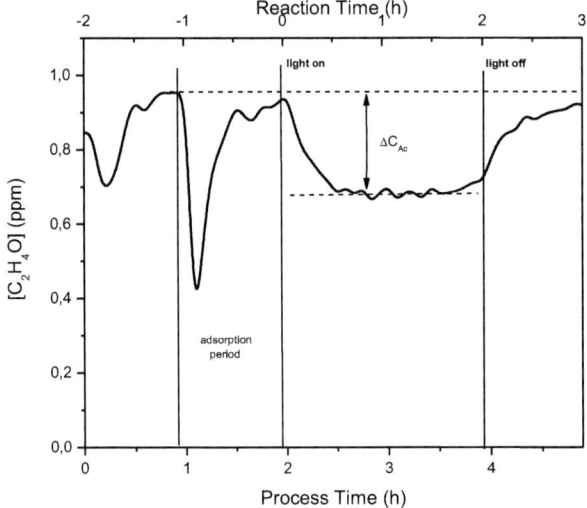

Figure IV-7. Comparison of 1ppm Acetaldehyde (—) degradation with 4 g P25 pressed powder photocatalyst applying 1 mW/cm² UV light photons without any filter.

Using the photocatalyst parameters for P25 presented under UV-A light irradiation a photonic efficiency of 0.15% can be calculated, with the values shown in Table IV-5.

Table IV-5. Acetaldehyde degradation results under UV-A light with P25 photocatalyst.

| Photocatalyst | (UV-light) | |
|---|---|---|
| | Acetaldehyde | |
| | Rate [mol/s] | Intensity [mol/s] |
| P25 (4.0g) | $1.89*10^{-10}$ | $1.19*10^{-7}$ |
| % Photonic Efficiency | 0.15 | |

The acetaldehyde degradation in gas phase with P25 as photocatalyst using a Pilkington green filter, and a polycarbonate filter (Figure IV-8) was considered as a comparison test to observe if the S-TiO$_2$-Fe$^{3+}$ has a better performance than P25 under visible light. Compared to the S-dopedTiO$_2$-Fe$^{3+}$

## Results and Discussion

photocatalyst it is clearly lower the capacity of degradation of the considered Degussa P25 photocatalyst beneath the same conditions.

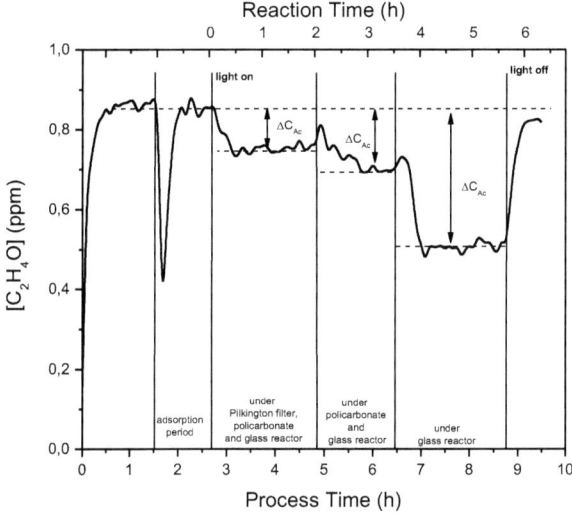

Figure IV-8. Comparison of 1 ppm Acetaldehyde (—) degradation with 4 g P25 pressed powder photocatalyst applying visible light photons under a Pilkington green filter, polycarbonate and without any filter.

The results of the experiments with the P25 photocatalyst by the decomposition of acetaldehyde under visible light are presented in Table IV-6.

Table IV-6. Acetaldehyde degradation results under visible light using different filters with P25 photocatalyst.

| Photocatalyst | Acetaldehyde (VIS-light) | | | | | |
|---|---|---|---|---|---|---|
| | Pilkington filter, Polycarbonate and glass reactor | | Polycarbonate and glass reactor | | glass reactor | |
| | Rate [mol/s] | Intensity [mol/s] | Rate [mol/s] | Intensity [mol/s] | Rate [mol/s] | Intensity [mol/s] |
| P25 (4.0g) | $0.732 \times 10^{-10}$ | $8.62 \times 10^{-8}$ | $1.11 \times 10^{-10}$ | $1.09 \times 10^{-7}$ | $2.37 \times 10^{-10}$ | $1.34 \times 10^{-7}$ |
| % Ph. Efficiency | 0.085 | | 0.102 | | 0.177 | |

# Results and Discussion

## IV.3. Durability of Ag-TiO$_2$ Photocatalysts Assessed for the Degradation of Dichloroacetic Acid

### IV.3.1. Preparation of Ag-TiO$_2$ and colloidal TiO$_2$ photocatalyst

The Ag-TiO$_2$ photocatalysts were prepared by a photodeposition process. Degussa P25 was used as the TiO$_2$ photocatalyst. P25 TiO$_2$ consists of 80% anatase and 20% rutile with an average primary particle diameter of 55 ± 5 nm and a BET surface area of 48 m$^2$/g.

Silver ions in the form of AgClO$_4$ were then added to the TiO$_2$ slurry at an initial pH of approximatley 6 to obtain the desired Ag$^+$:Ti ratio. The suspension was illuminated for 24 hours under a UV-A light intensity of 1 mW/cm$^2$. A series of Ag-TiO$_2$ photocatalysts with a respective Ag loading of 0.1, 0.2, 0.35, 0.5, 1 and 2 atom% was prepared. The samples were labeled as xAg-TiO$_2$ where x is the amount of Ag$^+$ added given in atom%.

Colloidal TiO$_2$ was prepared by a method described by Bahnemann [71] that involved the dropwise addition of titanium tetraisopropoxide dissolved in 99.99% 2-propanol to an aqueous hydrochloric acid solution of pH 1.5. The final concentrations of TiO$_2$ and 2-propanol were 1.5 x 10$^{-2}$ and 1.2 M, respectively. The reaction volume was 1L. This mixture was stirred overnight employing a magnetic stirrer. The final solution was optically transparent. Evaporation under vacuum was used to collect the TiO$_2$ powder which can be resuspended in water and other polar solvents (e.g. alcohols) to form transparent colloidal suspensions.

For the preparation of the Ag-TiO$_2$ photocatalysts, the first indication of the deposition of metallic silver onto the TiO$_2$ surface upon irradiation of the TiO$_2$ and Ag$^+$ suspensions, was a color change of the particles from white to brown. A higher concentration of silver ions corresponded to a higher intensity of this brownish color. According to the literature, the photodeposited Ag should be present in metallic form, based on XRD analysis of Ag/TiO$_2$ particles prepared by using 20.0 atom% Ag$^+$ loading which confirmed the presence of metallic silver [72]. At lower Ag content, metallic silver could not be detected since the mass of silver in the sample was below the detection limit of the employed instrument [72].

Results and Discussion

## IV.3.2. Analysis and characterization of the Ag-TiO$_2$ photocatalysts

TEM analyses of the photocatalysts were carried out with the aim to examine if any visible morphological changes had occurred to the photocatalyst samples after the second recycle. The TEM analysis was combined with an EDXS-REM from which elemental maps of Ti, Ag, and Cl were obtained to observe their distribution in the samples. As a blank comparison, Figure IV-9 presents an EDX spectrum of Degussa P25 before the photodeposition of silver or the photodegradation of DCA occured. Figure IV-10 shows the same sample after Ag was deposited on the surface. The results are presented for the 0.35 Ag-TiO$_2$ sample, which had not the highest initial activity at pH 3 but the greatest performance after 2 recycles. Figure IV-11 shows the EDX spectrum after the photodegradation of DCA in three consecutive runs. The presence of Cl$^-$ ions detected, indicative of the poisoning of the photocatalyst surface.

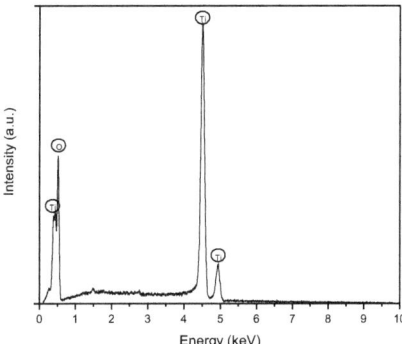

Figure IV-9. EDXS-spectrum of Degussa P25 prior to the photodeposition of silver or the photodegradation of DCA.

61

Results and Discussion

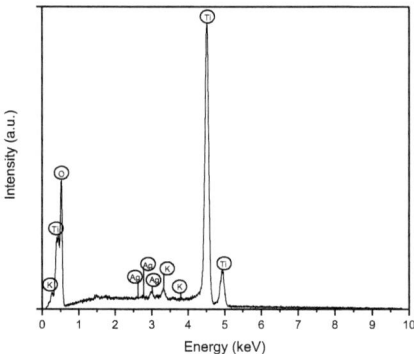

Figure IV-10. EDXS spectrum of 0.35 Ag-TiO$_2$ photocatalyst prior to the photodegradation of DCA.

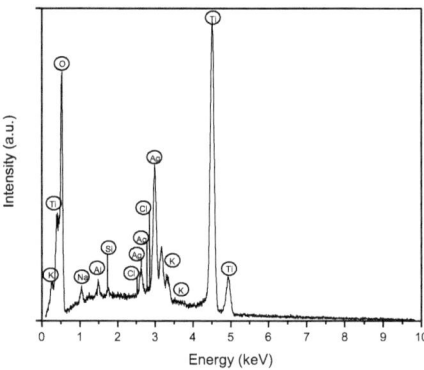

Figure IV-11. EDXS spectrum of 0.35 Ag-TiO$_2$ photocatalyst after photodegradation of DCA in three consecutive runs.

In Figure IV-12, the STEM image and the elemental maps of Ti and Ag in the 0.35 Ag-TiO$_2$ are shown. These figures show the distinct silver deposits which are present in the samples. The size of the silver deposits in 0.35 Ag-TiO$_2$ sample is larger than that typically presented in the literature [41]. This is believed to be due to the comparatively long irradiation time of 24 hours during the photodeposition reaction. Figure IV-13 shows the STEM image of the same 0.35 Ag-TiO$_2$ photocatalyst sample after the DCA photodegradation reaction. The corresponding x-ray maps of Ag and Cl confirm the presence of the Cl⁻ ions and the close association of Ag with the Cl⁻ ions.

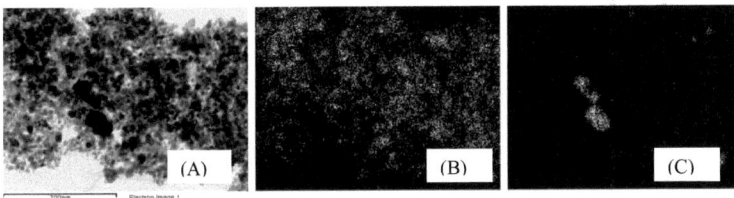

Figure IV-12. Electron microscopy analysis of 0.35 Ag-TiO$_2$ photocatalyst particles before the DCA photodegradation reaction (A) STEM image of particles, (B) corresponding Ti x-ray map and (C) corresponding Ag x-ray map.

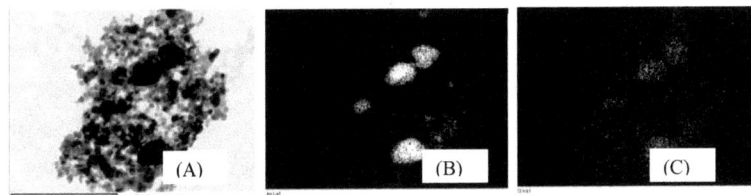

Figure IV-13. Electron microscopy analysis of 0.35 Ag-TiO$_2$ photocatalyst particles after the DCA photodegradation reaction (A) STEM image of particles, (B) corresponding Ag x-ray map (C) corresponding Cl x-ray map.

Based on the above electron microscopy results, it is clear that following the DCA degradation, there is a considerable amount of Cl⁻ ions on the photocatalysts surface.

It is believed that the chloride ions are merely physically adsorbed onto the surface of the photocatalyst rather than reacting with the silver deposits. This was supported by the fact that the photocatalysts were found to regain their activity by a simple washing procedure. When considering the preparation of the Ag-TiO$_2$ particles by photodeposition, the silver salt (AgClO$_4$) was added at slightly acidic pH (pH 6) and the titration to the reaction pH followed afterwards. Hence at pH 10, the silver ions are expected to stay adsorbed on the surface and there will be hardly any external formation of AgOH precipitates. Similarly at pH 3, the silver ions are expected to remain adsorbed onto the TiO$_2$ surface. This is believed to have minimized the reaction between Cl⁻ ions and Ag⁺ or AgOH as the photodegradation reaction proceeded.

If, on the other hand the titration of the titania to pH 3 or pH 10 was to be done before the silver was added, different results would have been expected. At pH 3, there would have been a lower amount of silver adsorbed onto the titania surface since both $Ag^+$ and the titania surface are positively charged. At pH 10, AgOH will most certainly form adjacent to the titania particles. When $Cl^-$ anions are formed in solution these are expected to react with the $Ag^+$ ions or AgOH to form the silver halides (silver chloride has a solubility of 0.8 ppm ($K_{sp}$ (AgCl) = 1.6 x $10^{-10}$) which is pH independent, so most of the silver chloride would have precipitated out). Upon UV exposure, the silver salts/ compounds are expected to be reduced to silver atoms (as in the photographic process). This would have quite a significant impact on the photolytic formation of the silver deposits on the titania surfaces and the subsequent photocatalytic behavior of the material.

It is also often reported that small photodeposited Ag particles grow due to Oswald ripening as the photodeposition reaction proceeds. For example, at a silver loading of 0.39 wt%, after 60 seconds of irradiation the size of the Ag deposits was 3 nm which increased as the irradiation time was doubled with some larger Ag deposits (23 nm) being formed [73]. TEM analysis was carried out on the 0.35 Ag-$TiO_2$ photocatalysts before and after recycling to check, if the silver deposits were in fact reforming after multiple exposures. The results presented in Figure IV-14 demonstrate that there was no observable change in the size of the Ag deposits. This was as expected since the preparation process allowed 24 hours of illumination for the deposition of the silver particles. As mentioned earlier, the initial Ag deposits were in fact much larger than Ag-$TiO_2$ photocatalysts which are typically studied. Such studies often employ much shorter irradiation times (30 minutes to few hours) [41]. For our case, the intention was to mimic real conditions which in fact can involve long periods of illumination as several pilot plant studies on photocatalysis have demonstrated [74,75]. Hence, the procedure for preparing the Ag-$TiO_2$ photocatalyst particles is critical in determining the nature of the Ag-$TiO_2$ particles that are formed as well as the subsequent changes and or interactions which may occur as the photocatalytic reaction proceeds under real conditions.

## Results and Discussion

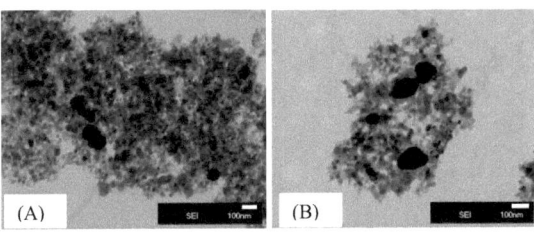

Figure IV-14. TEM images of 0.35 Ag-TiO$_2$ photocatalyst (A) before and (B) after recycle.

### IV.3.3. Degradation of DCA with photodeposited silver on Degussa P25

The effect of recycling the 0.35 Ag-TiO$_2$ photocatalyst on the photocatalytic degradation rate of DCA is shown in Figure IV-15 and once again in Figure IV-16 employing the washing technique in Figure IV-16. The reaction conditions were the same as described previously for the experiment with pure P25 TiO$_2$. As it can be seen from Figure IV-15, the reaction rate is found to be lowered after the first run, and is lowered even more during the 3$^{rd}$ run. Once again, the reaction rate is further reduced when this third run is carried out after the addition of 4 mM Cl$^-$. Moreover, the reaction seems to stop in this case after about 20% of the initially present DCA has been degraded. This is consistent with the respective TOC results shown in Figure IV-18 (vide infra).

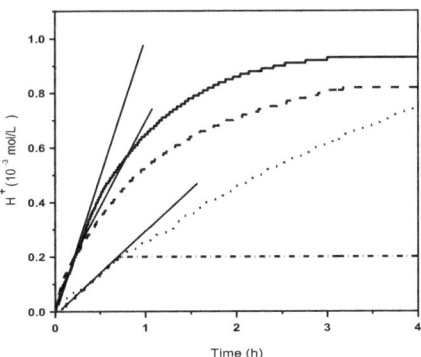

Figure IV-15. Degradation of DCA (shown as release of H$^+$) using the photocatalyst 0.35 Ag-TiO$_2$ at pH 3 in 3 consecutive runs, 1$^{st}$ run (—), 2$^{nd}$ run (---), 3$^{rd}$ run (···), a 3$^{rd}$ run (-·-) with addition of 4 mM Cl$^-$ before the run

## Results and Discussion

started and slope (—) used for the determination of the photonic efficiency of each run with $I \approx 3.39 \times 10^{-2}$ Einstein L$^{-1}$h$^{-1}$. The photocatalyst loading was 0.5 g/L.

In the next set of experiments, the aim was to investigate whether the negative effect of recycling the Ag-TiO$_2$ photocatalyst can also be avoided by a simple washing technique. Washing of the photocatalysts was carried out by recovering the material by centrifugation of the reaction slurry after each run and resuspending it in distilled water, followed by the appropriate adjustment of pH. The results are shown in Figure IV-16. As in the case of pure Degussa P25 it is obvious that this simple intermittent washing technique between each reaction run leads to a remarkable improvement in activity following the first and second recycles.

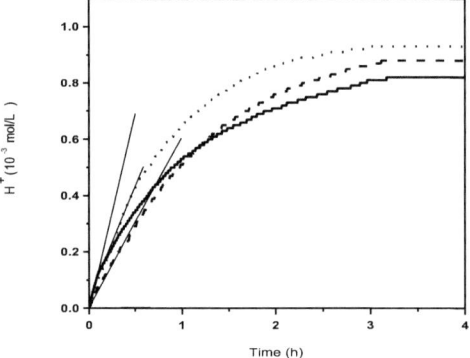

Figure IV-16. Degradation of DCA (shown as release of H$^+$) using the photocatalyst 0.35 Ag-TiO$_2$ at pH 3 in 3 consecutive runs with intermittent washing between runs, 1$^{st}$ run (—), 2$^{nd}$ run (---), 3$^{rd}$ run (···) and slope (—) used for the determination of the photonic efficiency with $I \approx 3.39 \times 10^{-2}$ Einstein L$^{-1}$h$^{-1}$. The photocatalyst loading was 0.5 g/L.

The results which are shown in Figures IV-3, IV-4, IV-15 and IV-16 clearly show that a "clean" surface is a prerequisite for the photocatalyst to exhibit its full activity independent of the absence or presence of co-catalysts such as Ag-deposits. The detrimental effect of co-dissolved ions such as Cl$^-$ is known since they are likely to adsorb onto the TiO$_2$ surface and can disturb the adsorption of the organic compounds [74].

Results and Discussion

### IV.3.4. Degradation of DCA with self prepared colloidal $TiO_2$ particles

The self prepared colloidal $TiO_2$ photocatalyst was tested with the same intention as pure Degussa P25 and P25-Ag using the identical experimental conditions. Figure IV-17 shows the results of these experiments. Obviously, the reaction rate and yield also decrease in subsequent runs. However, this poisoning effect appears to be much less pronounced for the colloidal particles as compared with pure or Ag-deposited Degussa P25. Even with the further addition of 4 mM of $Cl^-$ ions before the third run no substantial difference is detected, as compared with a third run without the addition of chloride ions. Thus the capacity of the self-prepared colloidal particles to accommodate chloride ions without any loss in photocatalytic activity appears to be very high.

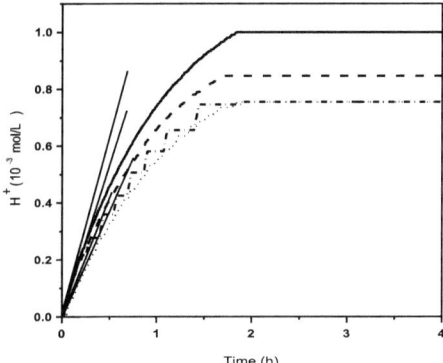

Figure IV-17. Degradation of DCA (shown as release of $H^+$) using the prepared colloidal $TiO_2$ photocatalyst at pH 3 in 3 consecutive runs with intermittent washing between runs, 1st run (—), 2nd run (- - -), 3rd run (···), a 3rd run (-··-) with addition of 4 mM $Cl^-$ before the run started and slope (—) used for the determination of the photonic efficiency of each run with $I \approx 3.39 \times 10^{-2}$ Einstein $L^{-1} h^{-1}$. The photocatalyst loading was 0.5 g/L.

### IV.3.5. Total organic carbon Ag-$TiO_2$ results

## Results and Discussion

The results of the total organic carbon (TOC) analyses performed at the end of each experimental run, i.e. after 4 hours of illumination and the H⁺ production efficiency in percent increased at the same time are presented in Figures IV-18 to IV-20. These figures provide a summary of the effect of recycling and washing, of the photocatalysts with respect to the mineralization of DCA. The TOC removal results are in fact in good agreement with the results obtained following the release of H⁺, evincing that the simple stoichiometry given in reaction (Eq. II.61) is indeed correct.

In Figure IV-18, the results for pure Degussa P25 are shown. The gradual decline in photoactivity following each recycle can be seen in set A for the consecutive runs. In set B, the strong detrimental effect of chloride ion addition before run 3 can be clearly seen. In set C, the photocatalyst was collected after each run by centrifugation, re-suspended in distilled water at the correct pH and placed back in the reactor. However, for the pure P25 photocatalyst the recovery of the photoactivity was not complete. Comparing the results obtained from set A with those of set C, however, it is obvious that the photoactivity after the washing technique is slightly higher than that observed without washing.

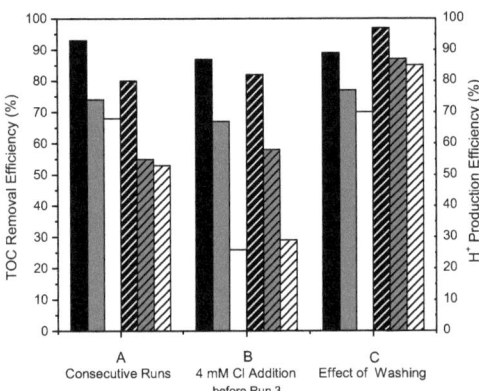

Figure IV-18. Removal of DCA (TOC removal) after 4 hours of illumination using the photocatalyst P25-TiO$_2$ at pH 3 in 3 consecutive runs (1$^{st}$ run ■, 2$^{nd}$ run ▨ and 3$^{rd}$ run □), the H⁺ production efficiency after 4 hours of illumination is also shown (1$^{st}$ run ▨, 2$^{nd}$ run ▨ and 3$^{rd}$ run ▨) in set A, with chloride ion addition before the third run (set B), and with intermittent washing between runs (set C). The photocatalyst loading was 0.5 g/L and the light intensity $I \approx 3.39 \times 10^{-2}$ Einstein L$^{-1}$h$^{-1}$.

## Results and Discussion

The results for the TOC removal using the 0.35 Ag-TiO$_2$ photocatalyst are presented in Figure IV-18. These are also in agreement with the DCA degradation results shown in Figures IV-15 and IV-16. The results of set C show the successful application of this simple washing technique resulting in an almost complete recovery of the photocatalytic activity. While the amount of DCA degraded after 4 hours of illumination even seems to slightly increase after each illumination/washing cycle (Figure IV-19, set C), the initial rate of degradation is found to decrease (Figure IV-16), in particular, for the second run. In comparison with the results obtained for pure Degussa P25, however, it is obvious that the photocatalytic activity of Ag-TiO$_2$ can be recovered much more richly by this simple washing procedure. As shown (of Figures V-12 and V-13), this might be due to the fact that the chloride ions show a strong tendency to adsorb at the silver clusters from where they can obviously be removed much easier than from the bare TiO$_2$ surface.

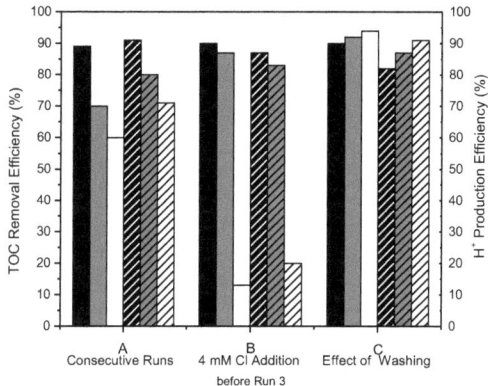

Figure IV-19. Removal of DCA (TOC removal) after 4 hours of illumination using the photocatalyst 0.35 Ag-TiO$_2$ at pH 3 in 3 consecutive runs (1$^{st}$ run ■, 2$^{nd}$ run ■ and 3$^{rd}$ run □), the H$^+$ production efficiency after 4 hours of illumination is also shown (1$^{st}$ run ▨, 2$^{nd}$ run ▨ and 3$^{rd}$ run ▨) in set A, with chloride ion addition in set B before the third run and with intermittent washing between runs (set C). The photocatalyst loading was 0.5 g/L and the light intensity $I \approx 3.39 \times 10^{-2}$ Einstein L$^{-1}$h$^{-1}$.

For the colloidal TiO$_2$ particles, the analysis of the TOC removal after 4 hours of illumination also showed the durability of this photocatalyst (the results are shown in Figure IV-20). In fact, this photocatalyst showed a greater resilience to deactivation compared to pure Degussa P25 and to the Ag-TiO$_2$ photocatalysts. This is believed to be due to the greater number of active sites provided by

## Results and Discussion

this photocatalyst for the photocatalytic reaction, hence minimizing the observed poisoning by the chloride ions. It is important to note that due to its colloidal nature, this photocatalyst could not be separated by centrifugation and hence it could not be washed between the recycles. Both, its advantages and drawbacks need to be taken into account when considering its application under real conditions.

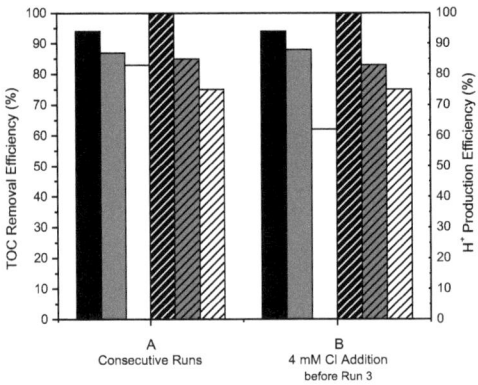

Figure IV-20. Removal DCA (TOC removal) after 4 hours of illumination using the colloidal-TiO$_2$ photocatalyst at pH 3 in 3 consecutive runs (1$^{st}$ run ■, 2$^{nd}$ run ■ and 3$^{rd}$ run □) the H$^+$ production efficiency after 4 hours of illumination is also shown (1$^{st}$ run ▨, 2$^{nd}$ run ▨ and 3$^{rd}$ run ▨) in set A and with chloride ion addition before the third run (set B). The photocatalyst loading was 0.5 g/L and the light intensity $I \approx 3.39 \times 10^{-2}$ Einstein L$^{-1}$h$^{-1}$.

In the next set of experiments, the effect of catalyst recycling on the photocatalytic DCA degradation rates was examined for a series of Ag-TiO$_2$ photocatalysts varying the amount of deposited Ag. These degradation experiments were carried out both under acidic (pH 3) and under basic (pH 10) conditions. The aim was to identify reaction conditions or photocatalyst composition where the observed reduction in photonic efficiency due to recycling can be minimized. Figure IV-21 provides a summary for the results obtained for the measured photonic efficiencies (based on the initial degradation rate of DCA as measured by the release of H$^+$).

## IV.3.6. Photonic efficiency Ag-TiO$_2$ results

In Figure IV-21, the photonic efficiency measured for each run is presented, comparing the fresh photocatalyst with the recycled ones after the second and third run. Figure IV-21 shows overall higher rates during the first run at pH 3 compared to pH 10 both, for pure Degussa P25 and when silver is photodeposited. This DCA degradation trend, although performed with a different photocatalyst, is in agreement with earlier studies reported by Bahnemann et al. [52]. They attributed the faster rates for a colloidal mixture of TiO$_2$-Fe with different Fe loadings under acidic conditions (pH 2.6) to the favorable, DCA coordination through the formation of bi-dentate complexes, which could not be formed in alkaline medium at pH 11.3. The absolute value of the photonic efficiency for the pure P25 photocatalyst obtained here is slightly lower than that reported previously [64], where a photonic efficiency $\xi$ = 3.9 % was measured at pH 3 under similar conditions. This could, for example, be explained by a variation of the activity of such a commercial product between different production batches which is well-known for most catalysts. It is interesting to note the influence of the chloride ions presence in the suspension. With the presence of 8 mM NaCl (2x2 mM for runs 1 and 2 plus 4 mM added before run 3) prior to run 3 the efficiency decreases by approximately 50 %, i.e. so the photocatalytic activity can be considerably inhibited by the presence of anions such as chloride. A similar yet less pronounced effect has been reported previously [64].

At pH 10, all the Ag-TiO$_2$ photocatalysts showed a similar performance regardless of the silver content. Given that P25 TiO$_2$ and Ag-TiO$_2$ photocatalysts had a similar photoactivity after recycling, this study demonstrates that at least for the photodegradation of DCA, there was no advantage in depositing Ag on the TiO$_2$ surface if the intention is to recycle the photocatalyst with no intermittent washing. For practical applications where recycling is part of a normal assumed procedure [75] this finding is of great consequence.

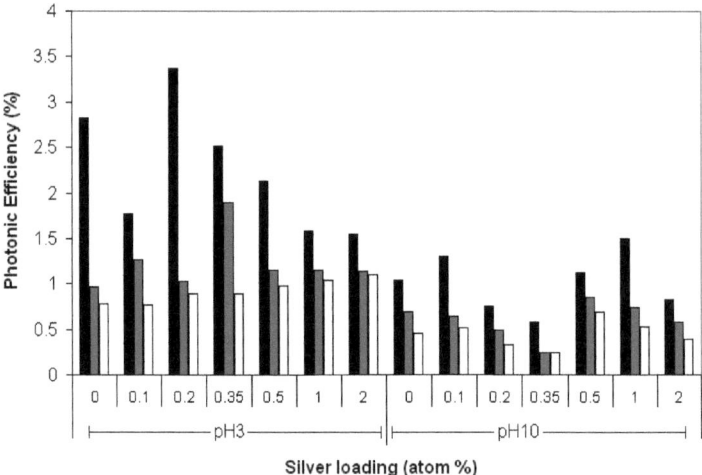

Figure IV-21. Effect of recycling Ag-TiO$_2$ photocatalysts on observed photonic efficiencies at pH 3 and pH 10 (1$^{st}$ run ■, 2$^{nd}$ run ■ and 3$^{rd}$ run □), degradation of 1mM DCA without any extra addition of chloride ions or washing technique performed. The photocatalyst concentration was 0.5 g/L with different silver loadings (atom%). The light intensity at pH 3 was $I \approx 3.39 \times 10^{-2}$ Einstein L$^{-1}$h$^{-1}$ and at pH 10 was $I \approx 3.74 \times 10^{-2}$ Einstein L$^{-1}$h$^{-1}$.

In Table IV-7, all photonic efficiency values are presented that have been calculated using equations (IV.1) and (IV.3) for each experiment. Previous reports ([64], Figure IV-17) concerning the photonic efficiency of the degradation of DCA in the presence of P25, are consistent with the photonic efficiency values obtained in this work. Here, the photonic efficiency was found to be between 2.83 and 3.69 % at pH 3 (cf. $\xi$ = 3.9 % at pH 3 [64]). Also, a decrease of the photonic efficiency was previously noted at pH 10 (cf. $\xi$ = 0.6 % at pH 10 [64]). The results obtained here concerning the influence of Cl$^-$ can be compared with a set of experiments carried out previously [64]. In this work the efficiency was reported to decrease from $\xi$ = 3.8 % with the number of subsequent experiments to $\xi$ = 2.0 % during run 3 and $\xi$ = 0.9% during run 6. Once again, the photonic efficiency recovered after washing of the photocatalyst (reaching $\xi$ = 2.2 % [64]) as confirmed in the present results.

## Results and Discussion

Table IV-7. Photonic efficiency values of the employed photocatalysts.
(Experiments as given in Figures IV-3, IV-4 and Figures IV-15 to IV-17).

| Photocatalyst [0.5 g/L] | | pH 3 | | | | | | pH 10 | | | | | | |
|---|---|---|---|---|---|---|---|---|---|---|---|---|---|---|
| | | Rate [mol/Lh] | | | Intensity [mol/Lh] 1.00E-04 | Photonic Efficiency (%) | | | Rate [mol/Lh] | | | Intensity [mol/Lh] 1.00E-04 | Photonic Efficiency (%) | | |
| | | Run 1 | Run 2 | Run 3 | | Run 1 | Run 2 | Run 3 | Run 1 | Run 2 | Run 3 | | Run 1 | Run 2 | Run 3 |
| P25 | | 9.58 | 3.26 | 2.65 | 3.39E-02 | 2.83 | 0.96 | 0.78 | 3.90 | 2.60 | 1.70 | 3.74E-02 | 1.04 | 0.70 | 0.45 |
| P25-Ag % | 0.1 | 6.00 | 4.30 | 2.60 | 3.39E-02 | 1.77 | 1.27 | 0.77 | 4.86 | 2.39 | 1.95 | 3.74E-02 | 1.30 | 0.64 | 0.52 |
| | 0.2 | 11.42 | 3.48 | 3.00 | 3.39E-02 | 3.37 | 1.03 | 0.89 | 2.84 | 1.84 | 1.24 | 3.74E-02 | 0.76 | 0.49 | 0.33 |
| | 0.35 | 8.50 | 6.40 | 3.00 | 3.39E-02 | 2.51 | 1.89 | 0.89 | 2.16 | 0.95 | 0.95 | 3.74E-02 | 0.58 | 0.25 | 0.25 |
| | 0.5 | 7.22 | 3.88 | 3.33 | 3.39E-02 | 2.13 | 1.15 | 0.98 | 4.20 | 3.20 | 2.59 | 3.74E-02 | 1.12 | 0.85 | 0.69 |
| | 1 | 5.35 | 3.90 | 3.53 | 3.39E-02 | 1.58 | 1.15 | 1.04 | 5.61 | 2.78 | 2.00 | 3.74E-02 | 1.50 | 0.74 | 0.53 |
| | 2 | 5.26 | 3.84 | 3.74 | 3.39E-02 | 1.55 | 1.13 | 1.10 | 3.12 | 2.16 | 1.48 | 3.74E-02 | 0.83 | 0.58 | 0.40 |
| 0.35 (4 mM Cl) | | 8.03 | 5.45 | 2.35 | 3.39E-02 | 2.37 | 1.61 | 0.69 | | | | | | | |
| 0.35 Washed | | 12.50 | 6.17 | 8.54 | 3.39E-02 | 3.69 | 1.82 | 2.52 | | | | | | | |
| P25 before 3rd run (4 mM Cl) | | 9.32 | 3.14 | 0.75 | 3.39E-02 | 2.75 | 0.93 | 0.22 | | | | | | | |
| P25 Washed | | 12.30 | 9.37 | 9.31 | 3.39E-02 | 3.63 | 2.77 | 2.75 | | | | | | | |
| TiO$_2$ colloid | | 11.40 | 9.07 | 6.73 | 3.39E-02 | 3.37 | 2.68 | 1.99 | | | | | | | |
| TiO$_2$ colloid before 3rd run (4 mM Cl) | | 11.20 | 8.92 | 6.44 | 3.39E-02 | 3.31 | 2.63 | 1.90 | | | | | | | |

A closer inspection of the results shown in Figure IV-21 reveals that while at pH 10 there is no clear relationship between the observed photonic efficiency and the absence or presence of Ag on the photocatalyst, at pH 3 the fresh 0.2 Ag-TiO$_2$ photocatalyst slightly outperforms pure Degussa P25. For the second run, however, the situation has changed with 0.35 Ag-TiO$_2$ now being the best photocatalyst for the DCA degradation. Finally in the third run of all tested catalysts exhibit almost the same photocatalytic activity ($\xi \leq 1$ %). Apparently, the Cl$^-$ ions formed during the photocatalytic oxidation of dichloroacetate are able to completely "neutralize" any enhancement in activity induced by the Ag clusters. Based on the experimental evidence given in Figures IV-12 and IV-13, this is explained by the preferred adsorption of chloride ions on these silver clusters.

## Results and Discussion

In Figure IV-22, a summary is provided for the results obtained for the measured TOC removal after 4 hours of illumination for the same experiments shown in Figure IV-21. Figure IV-22 shows the TOC removal measured for each run, comparing the fresh catalysts with the recycled photocatalysts after the third run. Again, the results here show the decline in activity on recycling, with no particular trend with regard to the amount of silver deposited or to the reaction pH. It is clear from both, the results in Figures IV-21 and IV-22, that while there is an optimum silver loading at 0.2 atom% at pH 3, this 0.2 Ag-TiO$_2$ photocatalyst has no clear advantage compared to the other Ag-TiO$_2$ samples after recycling. Finally, it should be noted that while the initial degradation rate (used to calculate the $\xi$ values presented in Figure IV-21) seems to be rather sensitive to the presence of the Ag-clusters (at least at pH 3), there is no such dependence observed for the overall reaction yield after 4 hours of illumination (cf. Figure IV-22).

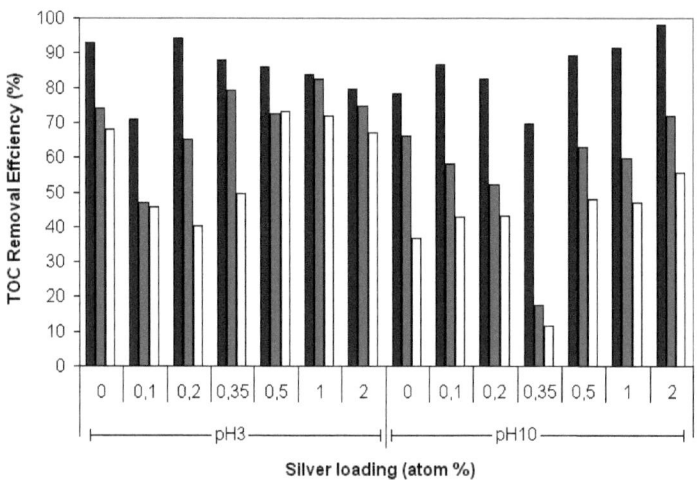

Figure IV-22. Effect of recycling Ag-TiO$_2$ photocatalysts on observed TOC removal at pH 3 and pH 10 (1st run ■, 2nd run ■ and 3rd run □), degradation of 1 mM DCA without any extra addition of chloride ions or washing technique performed. The light intensity at pH 3 was $I \approx 3.39 \times 10^{-2}$ Einstein L$^{-1}$h$^{-1}$ and at pH 10 was $I \approx 3.74 \times 10^{-2}$ Einstein L$^{-1}$h$^{-1}$.

## IV.3.7. Conclusions

In this study, it was possible to demonstrate the drop in photocatalytic activity after 3 consecutive runs for the degradation of DCA, using Ag-TiO$_2$ photocatalysts under both acidic and basic conditions. The aim was to test these photocatalysts under realistic process conditions which may involve photocatalyst recycle, and extended periods of illumination (up to 12 hours). Interestingly, a minimal reduction in activity was observed for high surface area, colloidal TiO$_2$. This is believed to be due to the greater number of active sites in this sample which made it more resilient to poisoning by the released chloride ions during the photodegradation of DCA.

The photocatalytic activity was easily recovered by a simple washing technique, which involved the collection of the photocatalyst particles by centrifugation and their re-suspension in a chloride free solution. The reversibility of the poisoning by the chloride ions gives evidence to support that the recycling of Ag-P25 TiO$_2$ photocatalysts (for a total 12 hours of illumination) does not have a permanent negative effect on its photocatalytic performance for the degradation of DCA. This was clearly affected by the choice of preparation procedure which minimized the interaction of the chloride ions with free Ag$^+$ ions or AgOH precipitates in solution, hence avoiding the formation of AgCl. The formation of AgCl in the system would have lead to complex photolytic reactions, whereby the overall performance for the photodegradation of DCA would have been difficult to predict.

## IV.4. Photocatalytic activities under visible light by S-doped $TiO_2Fe^{3+}$ photocatalyst

### IV.4.1. Preparation of S-doped $TiO_2$-$Fe^{+3}$ photocatalyst

As part of a collaboration project the group of Professor Teruhisa Ohno in Japan provided the S-doped $TiO_2$ powder. The starting material was synthesized by previously reported methods [16,56,76]. Titanium isopropoxide (50 g, 0.175 mol) was mixed with Thiourea (53.6 g, 0.70 mol) at a molar ratio of 1:4 in ethanol (500 ml). The solution was stirred at room temperature for 1 hour and concentrated under reduced pressure. After evaporation of ethanol, white slurry was obtained. The slurry was kept for 2 days at room temperature and a white powder was obtained. This powder was calcinated at various temperatures under aerated conditions, and yellow powder was obtained. In the S-doped $TiO_2$ powder no C or N atoms were included. An appropriate amount of $FeCl_3$ to obtain the absorption of 0.90 wt.% $Fe^{3+}$ ions was dissolved in deionized water, (300 ml) adsorbed. Three g of the doped $TiO_2$ powder were suspended in an $FeCl_3$ aqueous solution, and the solution was stirred vigorously for 2 hours After filtration of the solution the amount of $Fe^{3+}$ ions that remained was determined by UV absorption spectra; to estimate the amount of Fe ions adsorbed on the doped $TiO_2$ powder. The residue was washed with deionized water several times until pH of the filtrate was neutralized. The powders were dried under reduced pressure at 60°C for 12 hours.

### IV.4.2. Analysis and characterization of the S-doped $TiO_2$-$Fe^{3+}$ photocatalyst

Done in this work was the analysis and characterization of this S-$TiO_2$-$Fe^{3+}$ photocatalyst and were also realized photocatalytical experiments. The sulfur doped-$TiO_2$ material was characterized with an elemental analysis, diffuse reflectance and XRD spectra of S-doped $TiO_2$ loaded with $Fe^{3+}$ and tested by the degradation of DCA as a visible-light active photocatalyst in aqueous phase and by the degradation of $NO_x$ or acetaldehyde in the gas phase. The stability and durability of this material was also examined.

## Results and Discussion

The elemental composition of the photocatalyst is shown in Table IV-8. The presence of iron and sulfur can be confirmed in the powder. Some traces of nickel and copper are also present. The surface area of the S-doped $TiO_2$ with adsorbed $Fe^{3+}$ sample is 82.5 $m^2g^{-1}$.

Table IV-8. Elemental analysis of the S-doped $TiO_2$ with adsorbed $Fe^{3+}$ photocatalyst powder.

| Material | Element content in weight % | | | | | |
|---|---|---|---|---|---|---|
| | Ti | S | Fe | Ni | Cu | C |
| S-$TiO_2$-$Fe^{3+}$ | 96.72 | 2.4 | 0.68 | 0.11 | 0.09 | 0 |

On the surface of the yellowish S-doped powder photocatalyst was analyzed by XRD (Figure IV-23) in order to assure the structure of the doped $TiO_2$. According to XRD analysis the structure of the material corresponds to the anatase form of titanium dioxide. In Figure IV-23 (B) a small peak is noted showing the presence of a doped S $TiO_2$ crystalline material compared to a pristine anatase sample PC500 from Millenium in Figure IV-23 (A).

The XRD and reflectance spectra of the material were observed to determine the particle size of the doped sample. It was about 3.72 nm calculated by Debye-Sherrer's equation (IV.12).

$$D = \frac{0.9 \cdot \lambda}{\beta \cdot \cos\theta} \quad (IV.12)$$

$$\theta_m = \theta_2 - \theta_1 \quad (IV.13)$$

$$\beta = \frac{\theta_m}{2 \cdot 2\pi} \quad (IV.14)$$

$$\theta = \frac{\theta_h}{2 \cdot 2\pi} \quad (IV.15)$$

With:
$\lambda$ =0.154 [nm] wavelength of the X-ray; $\beta$ = full width at half maximum (radian measure); $\theta$ = Bragg angle; $\theta_m$ =width of the peak; $\theta_1$ = initial position of the peak at half maximum; $\theta_2$ =final position of the peak at half maximum; $\theta_h$ = highest position of the peak.

## Results and Discussion

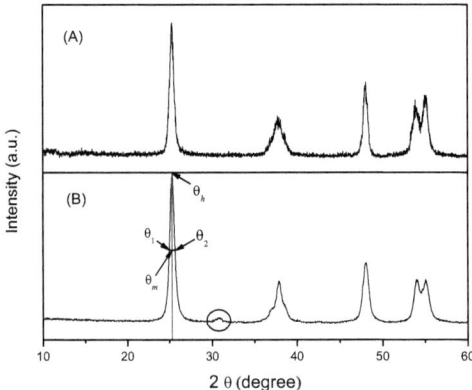

Figure IV-23. XRD of PC500 from Millenium pure anatase (A) and S-doped TiO$_2$-F$^{3+}$ (B) material.

The optical absorptivity of the nanoparticle powders is usually determined from the diffuse reflectance measurements, but in this case using the UV-Vis equipment can be directly obtained the reflectance data by the Kubelka-Munk function Figure IV-24 (A), as given in the following equation, where $R$ is the diffuse reflectance in equation (IV.16):

$$F(R) = \frac{(1-R)^2}{2R} \qquad (IV.16)$$

The normalized diffuse reflectance spectra presented as a Kubelka-Munk function of S-doped TiO$_2$ with iron ions adsorbed sample in Figure IV-24B presents a red shift shoulder between 380 and 500 nm of the absorption edge compared to the commercial Millenium PC500, bare anatase TiO$_2$. This result suggests a capacity of the material for absorption in the visible region.

Figure IV-25 shows the Tauc Plot [77] $((F(R) \cdot h\nu)^n \, versus \, h\nu)$ constructed from Figure IV-24A, which is used to determine the semiconductor band gap. If $n$ is taken as 0.5, the construction of the Tauc Plot leads to a linear Tauc Region which can be extrapolated to the photon energy axis to yield the semiconductor band gap. Degussa P25 powder is 3.22 eV, which agrees with other reports [78-80].

## Results and Discussion

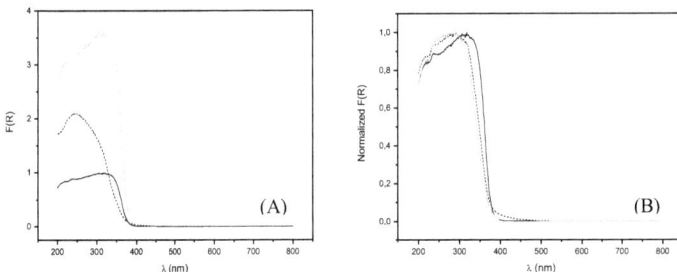

Figure IV-24. Reflectance function (A) and normalized Reflectance function (B) of P25 (—), anatase PC500 (—) and S-doped $TiO_2$-$Fe^{3+}$ (---).

The band gap of the S-doped $TiO_2$-$Fe^{3+}$ material 3.35 eV was obtained by the reflectance function for an indirect semiconductor [6] (Figure IV-25). Anatase was the $TiO_2$ structure confirmed by XRD of the doped photocatalyst (Figure IV-23), what resulted consistent with the band gap value obtained. But interestingly a blue shift shoulder appeared opposite to the red shift shoulder presented in Figure IV-24B. These red-blue shift shoulders could represent a capacity of the photocatalyst to absorb visible light not depending on the mathematical treatment, which could be applied to the data. Through these mathematical functions was easier to observe, what in Figure IV-24A just if carefully can be appreciated as a weak presence of a red shoulder.

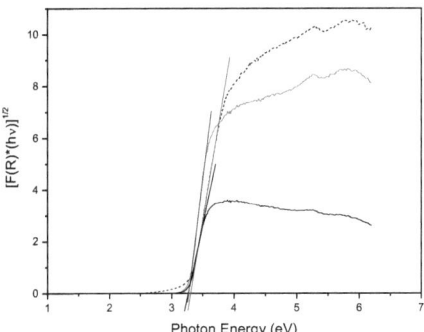

Figure IV-25. The band gap of P25 (—) and anatase PC500 (—) is 3.22 eV but for S-doped $TiO_2$-$Fe^{3+}$ is 3.35 eV (---).

79

## Results and Discussion

Using the band gap energy of the S-doped powder it was possible to make further calculations applying the Bruce equation (IV.17) for the even more approximated value of the material particle size of 4.1 nm. These calculations were made using values for $m_e^*$ and $m_h^*$ as determined by [81].

$$E(R) \cong E_g + \frac{\hbar^2 \cdot \pi^2}{2 \cdot R^2}\left(\frac{1}{m_e^*} + \frac{1}{m_h^*}\right) - \frac{1.8 \cdot e^2}{4 \cdot \pi \cdot \varepsilon \cdot \varepsilon_0 \cdot R} + smaller\ terms \quad \text{(IV.17)}$$

With:
$\lambda$ = 330nm; $E_R$ = excited state energy (6.02·10$^{-19}$ J); $E_g$ = band gap energy (5.37·10$^{-19}$ J); $\hbar$ = reduced Planck constant (1.054·10$^{-34}$ J-s); $\pi$ = 3.1416; $m_e^*$ = effective electro mass (0.3$m_e$); $m_h^*$ = effective hole mass (0.8$m_e$); $m_e$ = electron rest mass (9.1096·10$^{-31}$ kg); 2R = particle diameter (4.1 nm); e = electron charge (1.602*10$^{-19}$ C); $\varepsilon_0$ = dielectric constant (8.854·10$^{-12}$ F m$^{-1}$); $\varepsilon$ = TiO$_2$ anatase dielectric constant (31).

Considering the discrepancy between the values obtained in the particles size by XRD 20 nm and the calculated with the Bruce equation 4.1 nm, it was necessary to measure the particle size indirectly with a Dispersion Technology Inc. equipment for calculating particle size and the zeta point of charge of the powder to find a possible relation in-between.

Figure IV-26 presents the results of a zero point of charge study, considering a determined concentration of photocatalyst dissolved in distilled water at different pH values.

The zero point of charge, for this S-doped photocatalyst was determined at pH 6.76. This evaluation depends on the change of the polarity charge on the surface of the particles, which is obtained by measuring the electric potential in the suspension.

The suspension at pH 3 resulted extremely stable for even more than 24 hours after being stirred for two hours. This is explained by the potential value of the positive charged particles, which corresponds to higher interactions between the H$^+$ ion concentration in the suspension generating very stable suspended 40 nm crystalline particle clusters. The change in the pH promotes also a change in the formed clusters size, what means by increasing the pH value increases the cluster size, which is also reported here in Figure IV-26 showing also a direct relation to the potential measured. By increasing the pH value a regular potential behavior was observed. The potential curve decreases slowly as the pH value increases, even after crossing the cero point of charge.

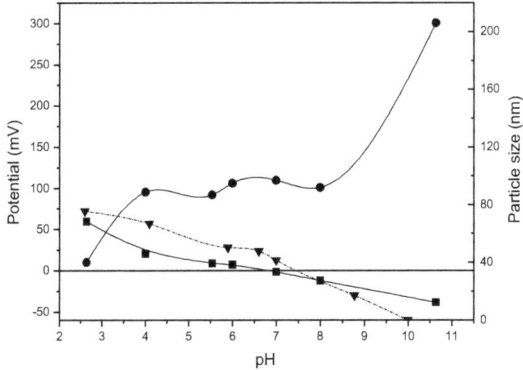

Figure IV-26. Zero point of charge comparison measurements of (5.0g/L) P25 (—▼—) and S-TiO$_2$-Fe$^{3+}$ (—■—) but also particle size (—●—) of S-TiO$_2$-Fe$^{3+}$ at different pH values in aqueous phase suspensions.

TEM analysis was carried out on the S-doped TiO$_2$-Fe$^{3+}$ material (Figure IV-27) to check the particle size and try to find if it is composed by crystalline structure. The powder seems to have a crystalline structure composed from different particle sizes from around 30 nm to less than 10 nm.

Figure IV-27. TEM image of S-doped TiO$_2$-Fe$^{3+}$ photocatalyst after preparation.

# Results and Discussion

In Figure IV-28 is presented the structure of the S-doped $TiO_2$ material. It can be observed representatively the replacement of one sulfur atom instead of an oxygen one in the lattice of the $TiO_2$ structure.

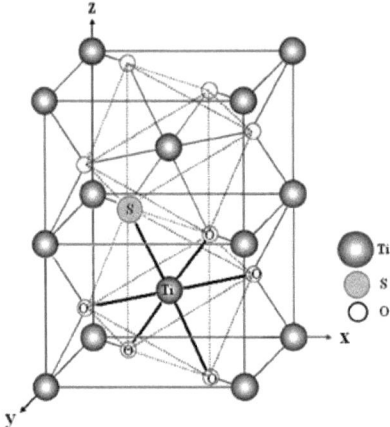

Figure IV-28. S-doped $TiO_2$-$Fe^{3+}$ photocatalyst.

### IV.4.3. Photocatalytic decomposition of DCA on S-doped $TiO_2$-$Fe^{3+}$ material

Experiments were carried out to measure and compare the photocalalytic activity under specific conditions. Figure IV-29 presents the experimental DCA degradation results with the S-doped $TiO_2$-$Fe^{3+}$ photocatalyst at two different concentrations under visible light ($\lambda$ > 420 nm). The lower concentration of photocatalyst 0.5 g/L shows essentially no decrease of TOC, with an initial value of 40 ppm and a value after eight hours of reaction of 38 ppm, while the reaction with the higher concentration 5.0 g/L resulted in a decrease in the concentration of almost 16 ppm. The chloride ions values are in agreement with the proportion of DCA decomposed. The curve describing the release of protons indicates that the reaction is slower with a lower concentration and it doesn't start immediately, what consequently causes a delay in the production of protons. This leads to the supposition that the photocatalyst has to be first under illumination for a period of time to be charged, so that the reaction can take place and start the release of protons. It looks like an induction period which seems to be longer for lower concentrations of photocatalysts. Usually the existence of an

## Results and Discussion

induction period as illustrated in Figure IV-29. means that the reaction initially proceeds so slowly that its progress is virtually unobservable. During this time, there is presumably a build-up towards a small but critical concentration of a reactive intermediate, which suddenly triggers off the main reaction. Induction periods are common in radical chain reactions, but in this case, were no intermediates observed. Last argument follows-up to consider that the initial period of time corresponds to a period for the photocatalyst to be energy loaded but not exactly an induction period.

The initial rate calculated by the initial concentration slope is 1.18 x10$^{-8}$ and 1.52 x10$^{-8}$ with a photocatalyst concentration of 0.5 g/L and 5.0g/L respectively. Between the rate values appears no large difference but is important to remark the dimension of the induction period required. Around one hour for the lower concentration instead of almost half an hour for the higher photocatalyst concentration. Due to this lack of energy with the lower photocatalyst concentration there is no significant degradation shown by comparing the values of TOC or Cl$^-$ ions. But with the 5.0 g/L photocatalyst concentration can be follow a degradation of the model compound DCA under visible light by measuring the released H$^+$, the TOC values and chloride ion concentration. All these results are in agreement with each other.

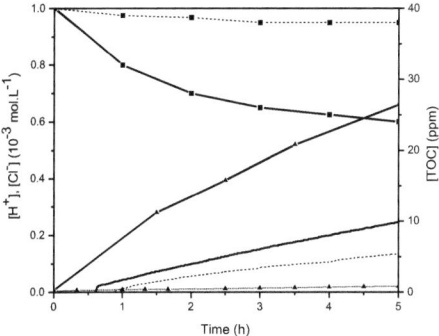

Figure IV-29. Decomposition of 1.7 mMol DCA under 0.896 mW/cm$^2$ visible light (420 nm cut-filter) at pH3 with S-doped TiO$_2$-Fe$^{3+}$ photocatalyst using 0.5 g/L and 5.0 g/L followed by the release or measurement of H$^+$(---); Cl$^-$(--▲--); TOC (--■--) and H$^+$(—); Cl$^-$(—▲—); TOC (—■—) respectively.

A similar experiment to the one described above was carried out under the same conditions but under UV-A light irradiation through a cut off filter of 320 nm. The experimental results are presented in Figure IV-30. Between the concentrations 0.5 g/L and 5.0 g/L of the S-doped TiO$_2$-Fe$^{3+}$ photocatalyst there was no major difference concerning the behavior during the photodegradation of

1.7 mMol of DCA. However, overall the photocatalyst under 320 nm is much more active than under a 420 nm wavelength cut off filter.

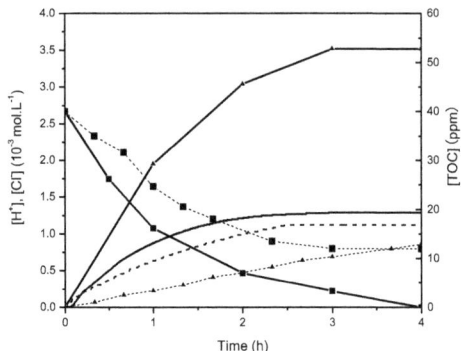

Figure IV-30. Decomposition of 1.7 mMol DCA under 30 mW/cm$^2$ UV-light (320 nm cut-filter) at pH3 with S-doped TiO$_2$-Fe$^{3+}$ photocatalyst using 0.5 g/L and 5.0 g/L followed by the release or measurement of H$^+$(---); Cl$^-$ (--▲--); TOC (--■--) and H$^+$(—); Cl$^-$(—▲—); TOC (—■—) respectively.

## IV.4.4. Degradation of DCA with S-doped TiO$_2$-Fe$^{3+}$ at pH 3 under visible light

For running a stability test by 1 mMol DCA degradation is necessary to follow the reaction for a longer period of time. For the analysis was decided to do all the experiments during 16 hours using a 420 nm cut off filter.

In Figure IV-31 is possible to follow a degradation of DCA by the release of H$^+$, the production of Cl$^-$ ions and the measurements of TOC. All this results confirm a not complete decomposition of 1mMol DCA for 15 hours of reaction but the intention is to test the stability of the material for a longer period of time under visible light energy. 1 mMol of DCA can be decomposed after 4 hours of reaction under UV-A reaction as presented in Figure IV-30.

Results and Discussion

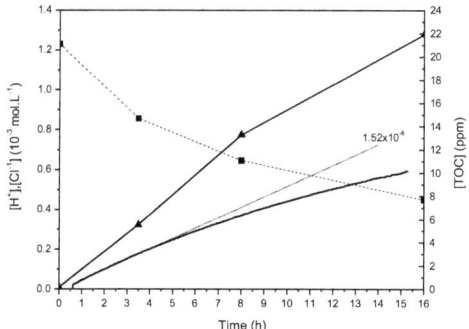

Figure IV-31. Photocatalytic activity by the degradation of 1 mMol DCA followed by the released of $H^+$ (—); the $Cl^-$ (—▲—) concentration and the TOC (--■--) measurements by using 5.0g/L S-doped $TiO_2$-$Fe^{3+}$ photocatalyst under 0.896 $mW/cm^2$ visible light 420 nm filter at pH3.

## IV.4.5. Different pH values for degradation of DCA with S-doped $TiO_2$-$Fe^{3+}$

The DCA degradation was evaluated at different pH conditions to test the photoactivity of the S-doped $TiO_2$ photocatalyst. The experiments were realized at pH 3; pH 7 and pH 9.

In Figure IV-32 is to observe which the better condition is for the better efficiency of the photocatalyst by the degradation of DCA. An acidic pH3 promotes a significantly higher decomposition of the DCA pollutant considering that the photocatalyst particles are positively charged, because of the $H^+$ concentration in the media. What implicates an easier adsorption interaction of the $DCA^-$ molecule on the particles and therefore its degradation.

# Results and Discussion

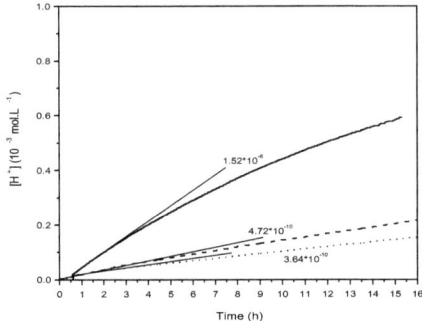

Figure IV-32. Photocatalytic activity degradation of 1 mMol DCA with 5.0 g/L S-doped $TiO_2$-$Fe^{3+}$ photocatalyst under 0.896 mW/cm² visible light (420 nm filter) by the $H^+$ release and compared at different pH conditions, pH 3 (—); pH 7 (---) and pH 9 (•••).

## IV.4.6. Stability test by a consecutively DCA degradation reactions with S-doped $TiO_2$-$Fe^{3+}$ at diverse pH conditions

As reported somewhere [63] a stability test for the photocatalysts can be performed by the consecutively DCA degradation runs to prove the capacity of the photocatalytical material to be reused for several times at different pH conditions.

In Figures IV-33 to IV-35 can be observed an increase in the reaction rate after the first degradation run. This might be for the consecutive 2nd and 3rd run because an induction period is not required any more for the photocatalyst to be energy loaded. The S-$TiO_2$-$Fe^{3+}$ photocatalyst can be interpreted as a stable material due to the continuous release of $H^+$ and its maintained kinetic slopes during the three consecutively DCA reaction runs at pH 3 (Figure IV-33).

Results and Discussion

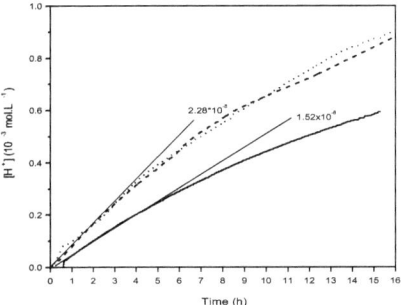

Figure IV-33. A consecutively repeated set of runs 1st run (—), 2nd run (---) and 3rd run (•••) was performed as a stability test for 5.0g/L S-doped $TiO_2$-$Fe^{3+}$ photocatalyst under 0.896 mW/cm² visible light (420 nm filter) at pH 3, during 16 hours each 1 mMol DCA degradation reaction, followed by the released of $H^+$.

Consecutively degradation runs of DCA with S-doped $TiO_2$-$Fe^{3+}$ at pH 7 were executed and the degradation rates obtained, are lower than the evaluated ones at pH 3. In Figure IV-34 are exhibited the pH 7 consecutive runs. The three of them present the same slope what enforces the stability condition of the material for a neutral pH value.

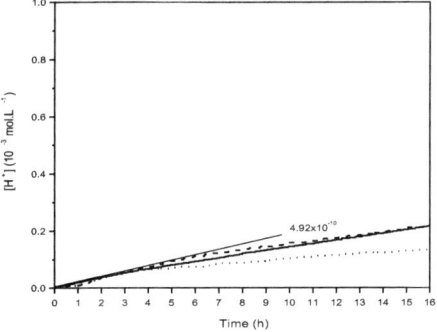

Figure IV-34. A consecutively repeated set of runs 1st run (—), 2nd run (---) and 3rd run (•••) was performed as a stability test for 5.0g/L S-doped $TiO_2$-$Fe^{3+}$ photocatalyst under 0.896 mW/cm² visible light (420 nm filter) at pH 7, during 16 hours each 1 mMol DCA degradation reaction, followed by the released of $H^+$ at pH 7.

Consecutively DCA degradation reactions with S-doped $TiO_2$-$Fe^{3+}$ at pH 9 are presented in Figure IV-35. The alkaline measured values determine a better condition for decomposing the DCA molecule compared to the neutral pH system, but just by a minor difference. Considering the different pH conditions studied, at pH 3 was obtained the highest conversion of DCA into $CO_2$ and $Cl^-$ ions related to the $H^+$ release.

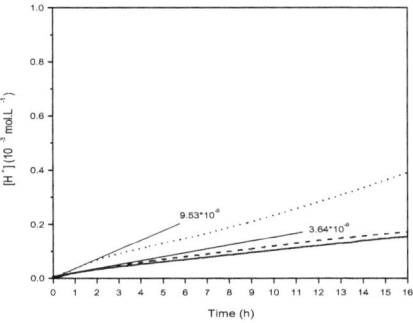

Figure IV-35. A consecutively repeated set of runs 1st run (—), 2nd run (---) and 3rd run (•••) was performed as a stability test for 5.0g/L S-doped $TiO_2$-$Fe^{3+}$ photocatalyst under 0.896 mW/cm² visible light (420 nm filter) at pH 9, during 16 hours each 1mMol DCA degradation reaction, followed by the released of $H^+$.

EDXS analysis was studied in order to have more information regarding the composition and stability of the sulphur doped photocatalyst before and after the consecutively DCA degradation reactions at pH 3 (Figure IV-36). The bare crystalline form of $TiO_2$ anatase (fresh) is considered as a blank for the comparison where no other peaks appear beside the $TiO_2$ ones. The (fresh) photocatalyst sample S-doped $TiO_2$ with adsorbed $Fe^{3+}$ ions before the DCA reaction, present the expected peaks of titan, sulphur and iron elements. An aluminum peak is also in the diagram but corresponds to the holder which was used to make sure the absence of carbon. One peak appears but it is attributed to the carbon adsorbed from the atmosphere as no carbon was present in the elemental analysis (Table IV-8). One chloride peak is observed but it is in accordance to the preparation method which disappears meanwhile runs were executed. After the first run the iron peak disappears what is logical because the iron ions were just adsorbed on the surface of the material however a small $Na^+$ peak appears from the titration system as an exchange ion to maintain the equilibrium charge on the surface due to the release of $Fe^{3+}$. After 32 hours continuous illumination of the second degradation reaction in the 2nd run the sulphur peak is almost disappeared of the crystalline structure. According to the result obtained from the third reaction (3rd run) one silicium peak is to be appreciated, which was gained

# Results and Discussion

from the quartz reactor material as a consequence of the continuously and vigorous stirring system. Beyond the silicium peak the pattern looks likewise to the pure anatase $TiO_2$ crystalline structure.

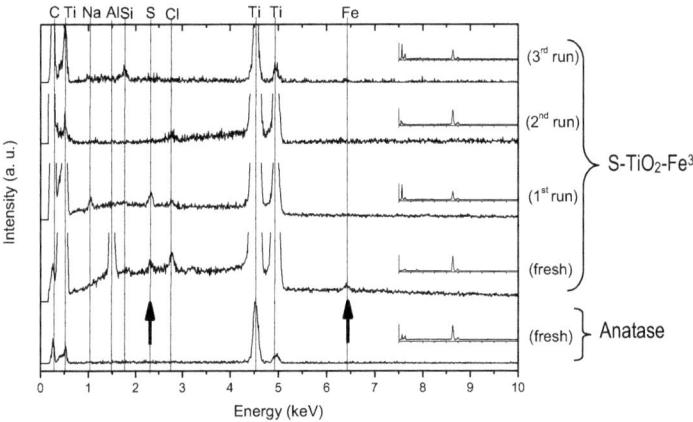

Figure IV-36. Comparison of 5.0g/L photocatalysts EDXS spectra of bare anatase and S-doped $TiO_2$-$Fe^{3+}$ before any reaction. And after stability test performed by the consecutively 1mMol DCA degradation runs with S-doped $TiO_2$-$Fe^{3+}$, 1st run, 2nd run and 3rd run under under 0.896 $mW/cm^2$ visible light, at pH 3. Figures inserted correspond to the complete spectrum of each analysis.

## IV.4.7. Comparison of commercial photocatalysts under UV-A and visible light

The photocatalytic activity of the S-doped $TiO_2$-$Fe^{3+}$ in Figure IV-37 was compared with two commercial photocatalysts Kronos VLP 7001 and Degussa P25. Degussa P25 was chosen as a standard photocatalyst. The Kronos VLP 7001 was tested because it is a photocatalyst material which is doped with 0.33 atom% sulfur and 0.9 atom% carbon, so it is expected to be active under visible light. The degradation of 1 mMol DCA was evaluated using 5.0 g/L photocatalyst in each case following the release of $H^+$. Degussa P25 still the most photocatalytic active material under UV-A light, and Kronos VLP 7001 was not as active as the S-doped $TiO_2$-$Fe^{3+}$ under the same illumination wavelength. But the most relevant result was the photocatalytic activity under visible light at $\lambda > 420$ nm presented by the S-doped $TiO_2$-$Fe^{3+}$ as the only photoactive photocatalyst of the compared ones.

# Results and Discussion

The photonic efficiency (see equation (IV.1)) in this case was calculated considering the incident photon flow as determined from the UV-A light meter measurements 30 (mW·cm$^{-2}$) by using a 320 nm cut off filter for the UV-A light and for the visible light experiments at 420 nm it was calculated (9.04*10$^{-1}$mW·cm$^{-2}$) from the xenon lamp spectrum (Figure IV-1) considering the sum of 7.5*10$^{-3}$ mW/cm$^2$ UV-A light measured and the calculated 8.96*10$^{-1}$ mW·cm$^{-2}$ visible light.

The irradiated surface area of the reactor was 8.042 cm$^2$ and the volume of the suspension 0.12 L.

The experimental results obtained from the comparison between the commercial photocatalysts and the sulfur doped in Figure IV-37 using a cut off filter of 320 nm are presented in Table IV-9.

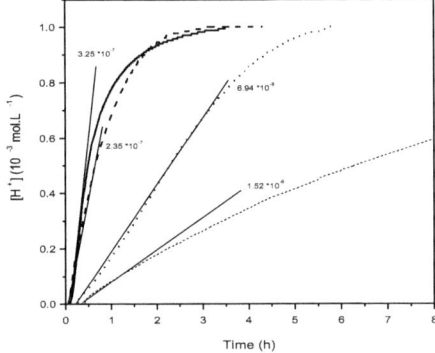

Figure IV-37. Photocatalytic activity comparison by the degradation of 1 mMol DCA between 5.0g/L of P25 (—), S-doped TiO$_2$-Fe$^{3+}$ (---) and Kronos (•••) photocatalysts under 30 mW/cm$^2$ UV-light (320 nm filter) resulting the S-doped TiO$_2$-Fe$^{3+}$ (···) as the only photoactive photocatalyst under 0.896 mW/cm$^2$ visible-light.(420 nm filter) at pH 3.

Table IV-9. Photocatalytic efficiency results of the comparison between the different commercial photocatalysts under 30 mW/cm$^2$ UV-A light (320 nm filter). (Figures IV-30 and IV-37)

| Photocatalysts | Rate (mol·L$^{-1}$s$^{-1}$) | Incident photon flux (mol·L$^{-1}$s$^{-1}$) | % of photonic efficiency |
|---|---|---|---|
| Degussa P25 (5.0g/L) | 3.25*10$^{-7}$ | 5.88*10$^{-6}$ | 5.52 |
| S-doped TiO$_2$-Fe$^{3+}$ (5.0g/L) | 2.35*10$^{-7}$ | | 3.99 |
| S-doped TiO$_2$-Fe$^{3+}$ (0.5 g/L) | 1.62*10$^{-7}$ | | 2.75 |
| Kronos VLP 7001(5.0g/L) | 6.94*10$^{-8}$ | | 1.18 |

# Results and Discussion

The photonic efficiency results of the different commercial photocatalysts are presented in Table IV-10. Is important to mention that no photoactivity was found by employing Degussa P25 or Kronos as photocatalysts under a 420 nm cut off filter.

Table IV-10. Photocatalytic efficiency results of the comparison between the different commercial photocatalysts under 0.896 mW/cm$^2$ visible light, using a filter $\lambda > 420$nm. (Figures IV-29 and IV-37)

| Photocatalysts | Rate (mol·L$^{-1}$s$^{-1}$) | UV-A light intensity (mW/cm$^2$) | Visible light intensity (mW/cm$^2$) | Total incident photon flux (mol·L$^{-1}$s$^{-1}$) | % of photonic efficiency |
|---|---|---|---|---|---|
| Degussa P25 (5.0g/L) | 0 | | | | 0 |
| S-doped TiO$_2$-Fe$^{3+}$ (5.0g/L) | 1.52*10$^{-8}$ | 7.5*10$^{-3}$ | 8.96*10$^{-1}$ | 2.127*10$^{-7}$ | 7.15 |
| S-doped TiO$_2$-Fe$^{3+}$ (0.5 g/L) | 1.18*10$^{-8}$ | | | | 5.55 |
| Kronos VLP 7001 (5.0g/L) | 0 | | | | 0 |

## IV.4.8. Photocatalytic decomposition of NO$_x$ or acetaldehyde on S-TiO$_2$-Fe$^{3+}$ nanoparticles under visible light in a gas phase reactor

With the described system in Figure III-2 was possible to follow the decomposition of NOx or acetaldehyde in gas phase.

Figure IV-38 shows the degradation of NO$_x$ experiment with S-doped TiO$_2$-Fe$^{3+}$ photocatalyst in gas phase under different wavelengths by applying a Pilkington green filter and a polycarbonate filter compared to the same system without any filter. The progress of the NO$_x$ removal implies complex parallel reactions which are not analyzed here in detail. With a luminescence detector can be follow the NO$_x$ degradation, the NO remotion and the produced NO$_2$ concentrations. The NO$_x$ and NO decomposition are the values considered for the photonic efficiency degradation comparison. The NO$_2$ produced is presented in the graph to confirm the degradation of the initial NO$_x$ compound but as these are directly related to each other is not necessary to present further photonic efficiency analysis concerning NO$_2$.

## Results and Discussion

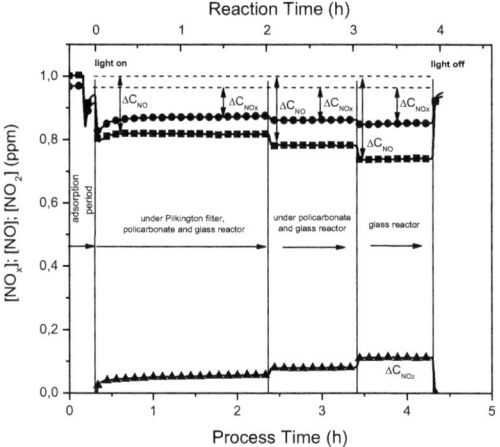

Figure IV-38. Comparison of 1 ppm NOx decomposition with 4 g S-doped $TiO_2$-$Fe^{3+}$ pressed powder photocatalyst applying visible light photons under a Pilkington green filter, polycarbonate and without any filter. The reaction was followed by measuring $NO_x$ (●); NO (■); and $NO_2$ (▲).

Photonic efficiency results of the $NO_x$ remotion under visible light using different filters with the S-doped $TiO_2$-$Fe^{3+}$ photocatalyst are presented in Table IV-11 and the respective results for the NO degradation in Table IV-12. After comparing the results is to observe a higher degradation efficiency by using photons of lower energy with the Pilkington and polycarbonate configuration filter. Which represent a robust confirmation of the visible light absorption capacity of the S-doped $TiO_2$ with $Fe^{3+}$ ions adsorbed.

Table IV-11. $NO_x$ degradation results under visible light using different filters with S-doped $TiO_2$-$Fe^{3+}$ photocatalyst.

| Photocatalyst | NOx (VIS-light) | | | | | |
|---|---|---|---|---|---|---|
| | Pilkington filter, Polycarbonate and glass reactor | | Polycarbonate and glass reactor | | glass reactor | |
| | Rate [mol/s] | Intensity [mol/s] | Rate [mol/s] | Intensity [mol/s] | Rate [mol/s] | Intensity [mol/s] |
| S-doped $TiO_2$-$Fe^{3+}$ (4.0g) | $1.89*10^{-10}$ | $8.62*10^{-8}$ | $2.07*10^{-10}$ | $1.09*10^{-7}$ | $2.28*10^{-10}$ | $1.34*10^{-7}$ |
| % Ph. Efficiency | 0.22 | | 0.19 | | 0.16 | |

## Results and Discussion

Table IV-12. NO degradation results under visible light using different filters with S-doped $TiO_2$-$Fe^{3+}$ photocatalyst.

| Photocatalyst | NO (VIS-light) | | | | | |
|---|---|---|---|---|---|---|
| | Pilkington filter, Polycarbonate and glass reactor | | Polycarbonate and glass reactor | | glass reactor | |
| | Rate [mol/s] | Intensity [mol/s] | Rate [mol/s] | Intensity [mol/s] | Rate [mol/s] | Intensity [mol/s] |
| S-doped $TiO_2$-$Fe^{3+}$ (4.0g) | $3.79*10^{-10}$ | $8.62*10^{-6}$ | $4.51*10^{-10}$ | $1.09*10^{-7}$ | $5.37*10^{-10}$ | $1.34*10^{-7}$ |
| % Ph. Efficiency | 0.441 | | 0.414 | | 0.401 | |

Another experiment was performed in gas phase to test the S-doped $TiO_2$-$Fe^{3+}$ photocatalyst remotion capacity by the degradation of acetaldehyde under a likewise NO degradation assembling with a Pilkington green filter, and polycarbonate filter below analog conditions. Figure IV-39 shows the degradation comparison between the filters configuration and the system without filter. Here is also possible to appreciate a degradation of the pollutant in the system which presents a considerably variation according to the applied filter. It can be observed the same trend as in the NO system by having a higher degradation without any filter but in this case also a higher photonic efficiency (Table IV-13).

In the acetaldehyde degradation experiment with Degussa P25 (see Figure IV-8) or with the S-doped $TiO_2$ (see Figure IV-39) is observed a very high acetaldehyde adsorption on the photocatalysts in the dark phase because the photocatalysts were preirradiated for three days so they still active ready to react adsorbing the model compound what is clearly noted before starting the degradation phase. It can be explained as a chain reaction process on the surface of the $TiO_2$ in presence of $O_2$ considering the initiation photocatalytic degradation steps (II.64 to II.69 ) [66].

# Results and Discussion

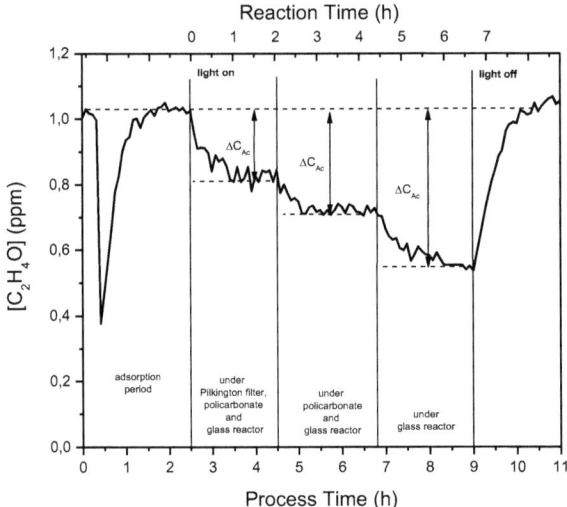

Figure IV-39. Comparison of 0.92 ppm Acetaldehyde (—) degradation with 4 g S-doped $TiO_2$-$Fe^{3+}$ pressed powder photocatalyst applying visible light photons under a Pilkington green filter, polycarbonate and without any filter.

Table IV-13. Acetaldehyde degradation results under visible light using different filters with S-doped $TiO_2$-$Fe^{3+}$ photocatalyst.

| Photocatalyst | Acetaldehyde (VIS-light) | | | | | |
|---|---|---|---|---|---|---|
| | Pilkington filter, Polycarbonate and glass reactor | | Polycarbonate and glass reactor | | glass reactor | |
| | Rate [mol/s] | Intensity [mol/s] | Rate [mol/s] | Intensity [mol/s] | Rate [mol/s] | Intensity [mol/s] |
| S-doped $TiO_2$-$Fe^{3+}$ (4.0g) | $1.5*10^{-10}$ | $8.62*10^{-8}$ | $2.2*10^{-10}$ | $1.09*10^{-7}$ | $3.28*10^{-10}$ | $1.34*10^{-7}$ |
| % Ph. Efficiency | 0.175 | | 0.202 | | 0.245 | |

## IV.4.9. Photonic efficiency S-$TiO_2$-$Fe^{3+}$ results

Figure IV-40 and Figure IV-41 summarize the decompositions rates and photonic efficiencies results respectively under visible light for the gas phase reactions, where the S-doped photocatalyst shows a

## Results and Discussion

higher rate and photonic efficiency than P25 with a lower photonic energy what can be considered as a convenient photocatalyst for decomposing pollutants under visible light.

By applying photons of higher energy (UV-A) all the reaction rates show an increasing trend what means that with the highest energy value applied both photocatalyst start to behave similar.

The sulfur doped photocatalyst can react better than P25 under photons of lower energy or in other words under visible light as can be supported with the photonic efficiencies results.

Comparing the decomposition of $NO_x$ and the acetaldehyde experiments is to be noted a considerable high adsorption on the photocatalysts surface only in the case of the acetaldehyde degradation system. During the period of adsorption in the darkness on the S-doped $TiO_2$-$Fe^{3+}$ photocatalyst were adsorbed 0.2097 ppm (Figure IV-39) almost double of the concentration adsorbed on the P25 photocatalyst 0.0989 ppm (Figure IV-8). This high adsorption phenomena can also induce to a better decomposition of the acetaldehyde compound.

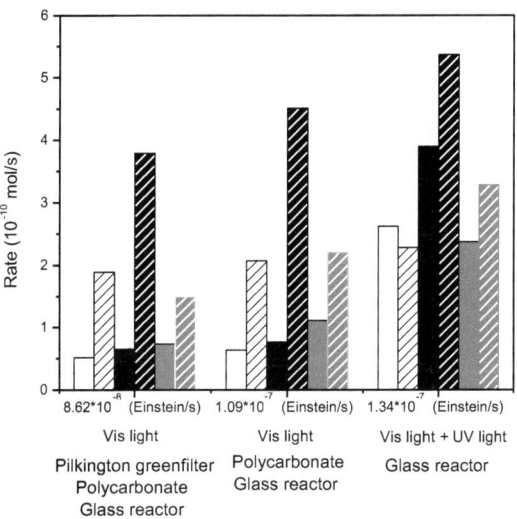

Figure IV-40. Comparison between the reaction rates decomposition of the gas phase systems applying the photocatalyst P25 for decomposing ($NO_x$□; NO■; acetaldehyde▨) and the photocatalyst S-$TiO_2$ ($NO_x$▨; NO▨; acetaldehyde ▨) under different intensities of visible light.

95

Results and Discussion

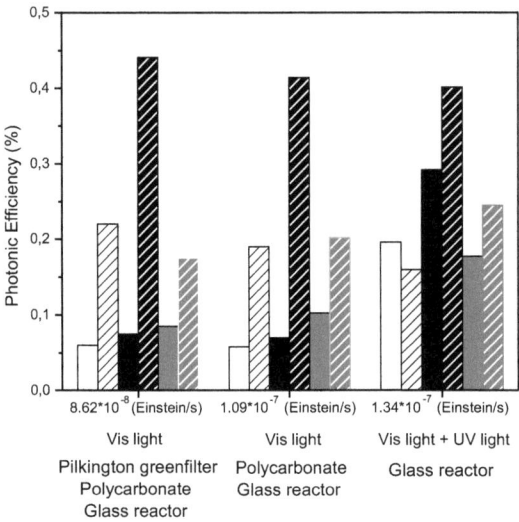

Figure IV-41. Comparison of 1ppm Acetaldehyde (—) degradation with 4 g P25 pressed powder photocatalyst applying visible light photons under a Pilkington green filter, polycarbonate and without any filter.

## IV.4.10. Conclusions

The aim of this work was to identify the best pH reaction conditions in aqueous phase; test the stability, durability; confirm the visible light energy absorption capacity of the S-doped $TiO_2$ with $Fe^{3+}$ ions adsorbed photocatalyst, by the degradation of different pollutants in aqueous and gas phase. The photonic efficiencies results obtained are based on the decomposition of the selected model compounds. The S-doped $TiO_2$-$Fe^{3+}$ photocatalyst resulted clearly active under visible light in all the tested cases.

The studied photocatalyst was stable under the different pH tested conditions pH 3; pH 7 and pH 9, where the best pH condition reaction result for the aqueous phase by the degradation of DCA was obtained at pH 3 which resulted stable for the first run.

After two consecutive DCA degradation runs is to see with an EDXS analysis a change in the crystalline structure. The S-doped photocatalyst material in aqueous phase after 32 hours of strength

## Results and Discussion

continuous reaction got released the sulfur from the anatase structure. The adsorbed $Fe^{3+}$ ions were released as well but these from the very beginning of the reaction. Never the less they took part of the photocatalytic reaction, as they were still present as oxidizing ions in the solution during the entire reaction. The zero point of charge was determined at pH 6.76.

In both gas phase inorganic and organic decomposition systems, $NO_X$ and acetaldehyde, is to observe, that without any filter is possible to have a higher conversion of the pollutants due to the major amount of photons available.

By comparing P25 and the S-doped $TiO_2$-$Fe^{3+}$ the highest photonic efficiency values are obtained by the sulfur doped $TiO_2$ photocatalyst in all the cases. Specially where the slice filters are used and the energy reached by the photocatalyst below the filters corresponds only to the visible light energy photons, which are taking part in the pollutant remotion, the difference between both photocatalysts is clearly high.

After the different analysis is widely confirmed the photocatalytic activity of the S-doped $TiO_2$-$Fe^{3+}$ photocatalyst under visible light.

## Results and Discussion

### IV.5. Solid state synthesis and characterization of $In_2Se_3$ nanoparticles deposited by heat treatment as a film electrode

Increasing attention has recently been given to indium chalcogenides due to their interesting electronic properties. These materials have been highlighted as having possible applications in solar cells[82]; $In_2S_3$ thin film solar cells have been reported with a 16.4% efficiency [83]. $In_2S_3$ has a reported band gap of 2.00–2.30 eV in the bulk[84]. However, there are reports of indium (I and III) sulfide films and particles with much larger band gaps. The preparation of thin films of the indium chalcogenides has been widely reported, in particular $In_2S_3$ which has been prepared by atomic layer deposition,[82] spray pyrolysis, chemical bath deposition, modulated flux deposition, and photochemical deposition. InS films are less well known and in most cases are amorphous. Annealing of these films leads to a complete transition to crystalline $In_2S_3$ [85]. By comparison there are relatively few reports in the literature of colloidal routes to indium chalcogenide nanomaterials. Nanoparticles of $In_2S_3$ have been synthesized by a variety of methods including, arrested precipitation yielding particles of between 2–3 nm [86], solvothermal reduction resulting in a range of sizes and morphologies and sonochemical synthesis yielding particles of 20–40 nm [84]. There is one report of InS nanoparticles in the literature [87]. The material was prepared by a single source route and was spherical. Indium selenide semiconductors have reported band gaps in the range from 1.73–2.09 eV [88]. Indium selenide thin films ($In_2Se$, InSe and $In_2Se_3$) are also well known [89-91]. The only report of a colloidal route to indium selenide nanoparticles details the photoelectrochemical reduction of an elemental selenium electrode in a solution of $InNO_3$ giving $In_2Se_3$ [92].

The literature on the structure of $In_2Se_3$ is confused and in some respects contradictory. Therefore only those features will be reproduced here on which there is reasonable agreement. Below 923 K the structure is composed of layers, which are weakly bound to each other. Each layer contains planes of selenium and indium in the sequence Se-In-Se-In-Se where the Se atoms in each plane build a triangular lattice with lattice constant a= 4.025 Å at RT.

Indium selenide (InSe) is a complex layered semiconductor with a direct optical band gap of 1.42 and an indirect band gap of 1.29 eV. However, InSe is a III–V lamellar semiconductor which meets the criteria for photo-intercalation. Depending on growth condition and on doping, this material exhibits n-

## Results and Discussion

or p-type behavior. InSe has two crystalline surfaces exhibiting very different physical properties. The cleavage surface perpendicular to the z-axis consists of selenium atom bounded together with covalent forces. The other surface parallel to the z-axis is made of selenium atom of adjacent layers being bounded by van der Waals forces. $In_2Se_3$ is of interest of its polymorphism and related metal-ion defect structure. It exhibits at least three different crystalline modifications denoted by $\alpha$, $\beta$ and $\gamma$ with transition temperatures of 200 and 650 °C, respectively, for the $\alpha-\beta$ and $\beta-\gamma$ transition.

$In_2Se_3$ is an n-type semiconductor in the form of hexagonal structure with a direct gap of 1.7 eV. The In–Se phase diagram shows four defined compounds at room temperature $\alpha$-$In_2Se_3$, $\alpha$-$In_3Se_8$, InSe and $In_2Se$. They are attractive semiconductors for potential applications in batteries and photo-electrochemical cells.

InSe correspond to a p-type semiconductor. From the literature is possible to compare different band gap values of the InSe or $In_2Se_3$ structures. This values vary from (InSe) 1.1 eV [93] to 2.23 eV [94] or 2.5 eV [95]. In our case for $In_2Se_3$ we obtained 2.74 eV for the indirect band gap and 3.2 eV for the direct band gap value. According to the Figure IV-51 (A) the synthesized $In_2Se_3$ fits better to an indirect semiconductor type.

One of the major difficulties in the photoelectrochemical energy conversion is the lack of stability of the semiconductor electrodes under illumination. Many attempts were made to stabilize the electrodes and to seek new electrode materials.

Layered compound semiconductors are expected to have high stability as photoelectrodes. Also because there is no dangling bond, which creates surface states, at cleaved surface of this materials, the surface recombination probability should be low that leads to high conversion efficiency. Although the photoelectrochemical characteristics of transition metal chalcogenides have been studied very extensively, those of III-VI compound semiconductors are not well known. Indium selenide (InSe) has an ideal energy gap (1.1-1.3 eV) for solar energy conversion devices but only a few reports are available for this application.

The possible use of this semiconductor compound as an absorber layer in solar cells requires its deposition on different substrates, with a satisfactory crystalline quality and reproducibility. The different substrates commonly used in photovoltaics are transparent conductive oxide (TCO) films for a window layer and metallic films (Au or Mo) for an ohmic back contact. One of the most common

## Results and Discussion

problems in the film growth processes is to effectively control the microstructure during the formation of a heterojunction. It is well known that the structural, electrical and optical properties of the material and thus the corresponding device performance are highly dependent on growth conditions as well as on heat treatment [96].

### IV.5.1. Synthesis development of $In_2Se_3$ and $In_6Se_7$ nanocrystals

Previously published procedures were tested for $In_xSe_y$ but the collected experience with those procedures produced so far extremely low yields [97] and in some cases did not even produce nanoparticles [98]. In the last two syntheses procedures, phosphines were used as reactants for example trioctylphosphine (TOP) and trioctylphosphine oxide (TOPO), but the use of phosphines should be avoided. Because In-TOP complexes thermally cleaves at the carbon phosphorus bond producing InP. But considering a physical property of selenium which is to be dissolved at high temperatures in some organic solvents was decided to follow a third synthesis route [99] which was reproduced and later modified to produce not only of $In_2Se_3$ but $In_6Se_7$ depending on the temperature and reaction period.

The synthesis for the $In_2Se_3$ nanoparticles was carried out as follows based on the synthesis of Tabernor *et. al.*, but with temperature modifications and longer reaction periods. Hexadecylamine (116 g) was degassed under reduced pressure at 140 °C for one hour and then heated to 300 °C in a sand bath under nitrogen atmosphere. Indium acetate (3 g, 10.2 mMol) and selenium (0.78 g, 10.2 mMol) were individually dispersed in a heated dodecylamine (15 mL) solution. Both dispersions were then simultaneously dropwise injected into the heated hexadecylamine. The reaction temperature dropped to approximately 280 °C. The reaction was heated again to 300 °C and the temperature maintained for 24 hours. The reactor was cooled quickly and after cooling, 100 mL methanol were added producing a mixture where the brown solid was isolated by centrifugation. It was also washed with approximately 10 mL chloroform and reprecipitated with acetone (100 mL). The synthesis procedure indicates a specific and detailed step by step preparation but the main reaction should take place as follows considering the synthesis of $In_2Se_3$ (Eq.IV.18) and respectively for the $In_6Se_7$ nanoparticles (Eq.19).

$$3CH_3-(R)-NH_2+2In(CH_3COO)_3+3Se \xrightarrow{\Delta}$$
$$In_2Se_3+3CO_2+3CH_3OO^-+3[CH_3NH_2-(R)]^+ \quad \text{(IV.18)}$$

$$9CH_3-(R)-NH_2+6In(CH_3COO)_3+7Se \xrightarrow{\Delta}$$
$$In_6Se_7+9CO_2+9CH_3OO^-+9[CH_3NH_2-(R)]^+ \quad \text{(IV.19)}$$

For the synthesis of $In_6Se_7$ nanoparticles in this case the period of reaction was longer but at lower temperature. It is suggested to heat $In_2Se_3$ up to 1000°C in an argon flow of 0.04 mL/ minute [100], what resulted in the slow loss of the more volatile component, selenium, enabling a very fine control of the stoichiometry of the sample to be maintained by varying the length of the heating period. In that case the X-ray powder analysis showed that the required period of reaction to produce $In_6Se_7$ was about 30 minutes. In this procedure hexadecylamine (50 g) was degassed under reduced pressure at 140 °C for one hour and then heated to 300 °C in an oil bath under nitrogen atmosphere. Indium acetate (1 g, 3.4 mMol) and selenium (0.268 g, 3.4 mMol) were individually dispersed in a heated dodecylamine (5 mL) solution. Both dispersions were then simultaneously dropwise injected into the heated hexadecylamine. The reaction temperature dropped to approximately 280 °C. The reaction was maintained for 72 hours at 100 °C. The reactor was cooled quickly and after cooling, 75 mL methanol were added producing a mixture where the solid was isolated by centrifugation. It was also washed with approximately 10 mL chloroform and reprecipitated with acetone (100 mL).

### IV.5.2. Analysis and characterization of the synthesized indium selenide material

The prepared indium selenide powder was analyzed by XRD (Figure IV-42). According to this analysis the structure of the material corresponds after the database WinXPOW and reported literature [99] to $In_2Se_3$. From the XRD pattern was determined the particle size 1.8 nm by the Sherrer equation (Eq. IV.12). It can also be observed a not complete crystalline structure but a low particle size by comparing the obtained wide peaks.

## Results and Discussion

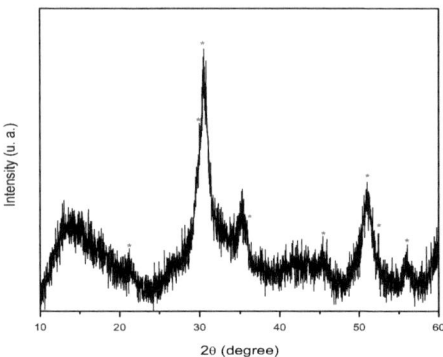

Figure IV-42. XRD pattern of the synthesized In$_2$Se$_3$ material (* peaks correspond to In$_2$Se$_3$ after WinXPOW database).

In Figure IV-43 and Figure IV-44, the In$_2$Se$_3$ SEM images are shown. The presented structure has a porous configuration which structure differs from the literature, where the synthesis products were In$_2$Se$_3$ wafers [99]. By controlling the temperature can be easily modified the stoichiometry of this compound but as can be observed, the structure results also a parameter which can be manipulated with specific organic precursors concentrations and conditions as a favorable media to induce a well mixed compound defining the formation of different structures.

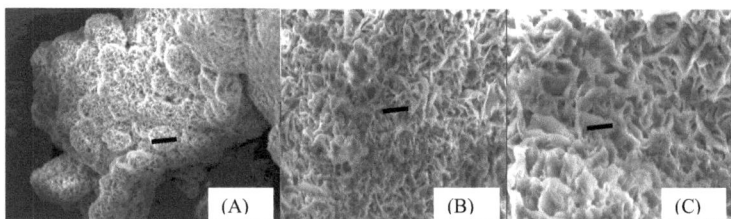

Figure IV-43. In$_2$Se$_3$ SEM images. The structure in (A) (—) represents 30 μm, in (B) (—) 20 μm and in (C) (—) 10 μm.

With a higher resolution in Figure IV-44, is possible to see on the surface of one slice, small particles attached to it. These homogeneously dispersed small particles appear to be around 10 nm (Figure IV-44) but after the XRD pattern (Figure IV-42) can be calculated smaller particles (1.8 nm) through the Scherrer equation. This discrepancy in the particle size is attributed to the reduced

## Results and Discussion

capacity of resolution of the scanning electrode microscope in presence of the amorphous organic material and for this reason only the biggest particles can be observed.

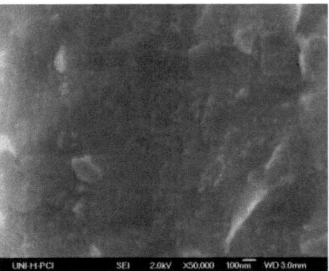

Figure IV-44. Surface of a $In_2Se_3$ slice.

During the $In_2Se_3$ synthesis is easy to loss selenium by the heat treatment as it can be removed by increasing the temperature. To confirm the presence of indium and selenium in the synthesis, the material was analyzed by EDXS before any further treatment (Figure IV-45). In deed was possible to find the indium and selenium elements with a higher amount of carbon due to the presence of hexadecylamine in the synthesis. It is also observed the presence of a small oxygen peak in the sample. Is important to follow the oxygen peak intensity to consider the presence of subproducts as $In_2O_3$, which is very easy to be formed by increasing the temperature, as was done here to obtain a higher crystalline structure in further heating treatments.

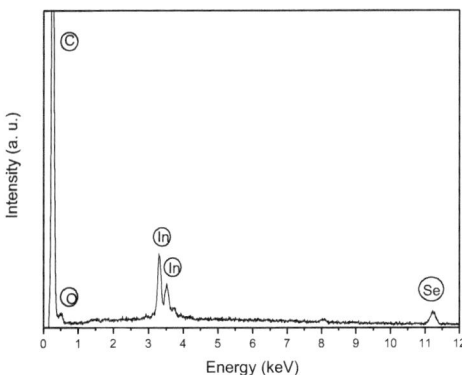

Figure IV-45. EDX spectrum of $In_2Se_3$.

## Results and Discussion

From a second synthesis with a longer period of reaction was produced $In_6Se_7$ as confirmed with the reported patterns from the WinXPOW database. The XRD pattern of the sample is presented in Figure IV-46. The particle size was (12.56 nm) obtained from the Sherrer equation (IV.12). This sample shows a higher crystalline structure than the one observed in the Figure IV-42 corresponding to the $In_2Se_3$. By comparing the different peaks can be proved the different stoichoimetry of both samples.

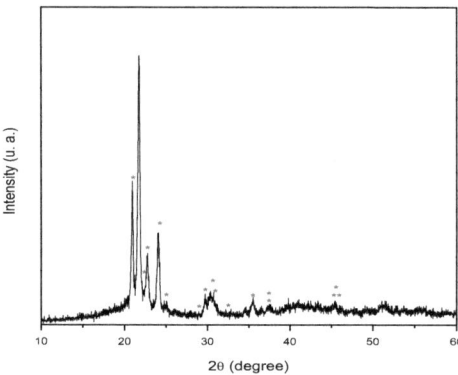

Figure IV-46. XRD pattern of the synthesized $In_6Se_7$ material (* peaks correspond to $In_6Se_7$ after WinXPOW database).

Scanning electron microscope images from the $In_6Se_7$ synthesis are presented in Figure IV-47. The material was found to be built of crystalline needles, they were in average 1 μm long. And there were individual spread needles observed but mostly the material agglomerated, it grew up producing big rocks due to the longer period of reaction (72 hours). Comparing the 1 μm length of the needles to the particle size calculated of the XRD-pattern from Figure IV-46 has to be considered a growing up of the needles through the smaller particles which should be present in a considerable higher quantity so that, through the X-ray pattern could be calculated the smaller particles. This also suggests the presence of the small crystalline structure particles building up the rocks.

## Results and Discussion

Figure IV-47. $In_6Se_7$ SEM image.

For this sample concerning the $In_6Se_7$ synthesis was also confirmed the presence of indium and selenium elements. The material was analyzed by EDXS (Figure IV-48). Through the peaks from the EDXS analysis can be observed a higher amount of selenium present in this sample. What fits with the relation of the indium element. In this case there is another arrangement in the structure which can be followed from the 1.5 eV and 11.3 eV selenium values, which are not appearing simultaneously in the $In_2Se_3$ sample. In the case of $In_2Se_3$ either one appears or the other depending on the concentration of selenium remaining in the sample. For lower selenium concentrations in the sample the 11.3 eV value disappears.

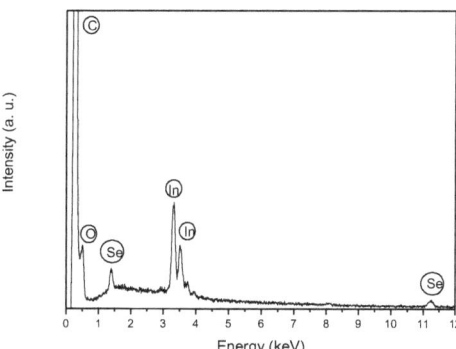

Figure IV-48. EDX spectrum of $In_6Se_7$.

The $In_2Se_3$ structure resulted more stable after the heating treatments than the $In_6Se_7$ crystals, therefore $In_2Se_3$ was chosen for further applications like the development of electrodes or experiments like the decomposition of $NO_x$ or acetaldehyde.

## Results and Discussion

The powder obtained from the In$_2$Se$_3$ synthesis after being heated in the oven for one hour at 500°C was analyzed by XRD (Figure IV-49). The material appears to be turned into a more crystalline structure but also seems to be changed into a mixture of In$_2$O$_3$ and In$_2$Se$_3$ by identifying the peaks in the WinXPOW database. The peaks at 22; 31; 46 and 51 degrees apparently corresponded to In$_2$O$_3$ but the rest of the peaks belonged to an In$_2$Se$_3$ phase [99]. After the Sherrer equation (IV.12) the particle size of the In$_2$Se$_3$ was 2.86 nm but for the In$_2$O$_3$ particle size corresponded 280 nm.

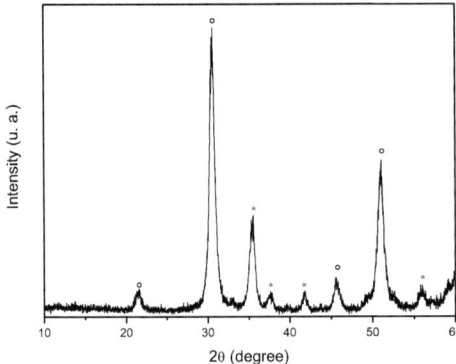

Figure IV-49. XRD of the synthesized In$_2$Se$_3$ powder after one hour of calcination at 500°C (* peaks correspond to In$_2$Se$_3$ and ° peaks to In$_2$O$_3$ after WinXPOW database)

The diffuse reflectance spectra (A) and the normalized diffuse reflectance spectra (B) as a Kubelka-Munk function of TiO$_2$ anatase PC500 and rutile R15 both from Millenium, were compared with the In$_2$Se$_3$ powder in Figure IV-50. In$_2$Se$_3$ shows a red shift of the absorption edge in relation to the bare anatase and rutile TiO$_2$. This result suggests a possible capacity of the synthesized In$_2$Se$_3$ material to absorb light in the visible region.

## Results and Discussion

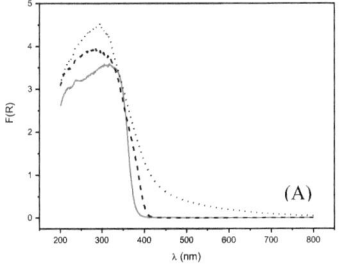
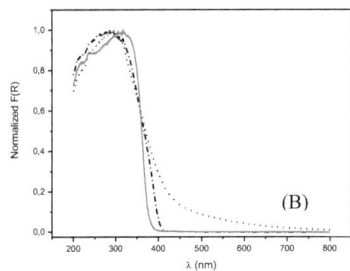

Figure IV-50. Reflectance function (A) and normalized Reflectance function (B) of anatase PC500 (—), rutile R15 (---) and In$_2$Se$_3$(•••).

To define the band gap energy of the In$_2$Se$_3$ material after the 500 °C heating treatment was used the diffuse reflectance spectra data (Figure IV-50A), which function was introduced in the form of the equation IV.20:

$$\alpha h \nu = [F(R) \cdot (h\nu)]^n \qquad (IV.20)$$

Where $R$ corresponds to the diffuse reflectance which is given as the Kubelka-Munk function obtained directly from the UV-Vis equipment measurement as $F(R)$ (Eq.IV.16), $h\nu$ is the photon energy, $\alpha$ is the absorption coefficient and $n$ depends on the nature of the transition. For direct transition $n$ =1/2 or 3/2, while for the indirect case $n$ = 2 or 3, depending on whether allowed or forbidden respectively.

With the obtained results from the In$_2$Se$_3$ powder reflectance spectra were calculated the indirect band gap 2.74 eV (Figure IV-51A) and the direct band gap 3.2 eV (Figure IV-51B). By comparing these gap values with the values presented in the literature, ranging from 1.84 eV to 2.09 eV [88] it can be considered that an indirect In$_2$Se$_3$ semiconductor type was synthesized. Confirming this presumption the tangent for 2.74 eV fits better (Figure IV-51 A), than the one presented in Figure IV-51 (B). Furthermore, it is the closest value to the literature range previosly mentioned. However, there are more discrepancies between the band gap values of this material in the literature, especially considering the reported indium selenide thin films, giving for example a band gap value of 2.5 eV [95]. This shows the difference between the literature and the obtained results. However it has to be mentioned, that the particle size is defining the band gap. As the particle size decreases, the

## Results and Discussion

absorption onset shifts further to the blue because of an additional quantum confinement in the x,y plane [98]. In this case the particles grow after the heating treatment at 500 °C and this is the reason for the shift in the band gap.

To define if this material corresponds to p or n-type semiconductor the most adeccuate procedure should be to have carried out some Hall measurements. As it was not possible, the research in the literature explained that apparently this material depends directly on the sinthesis procedure and even more to the annealing temperature applied. After Hall measurements by increasing the annealing temperature the n-type semiconductor changes to a p-type [88]. The last observation was supported with the argument that the nature of the n-type may be due to the presence of undiffused indium as impurity and this undiffused indium may act as donors. As annealing temperature was increased diffusion of indium may be more effective into the grain what may lead to p-type conductivity. According to this information and considering 500 °C as the annealing temperature the synthesized $In_2Se_3$ corresponds to n-indirect semiconductor type. To reinforce this information can be observed the positive inclination slope in the Mott-Schottky plot in Figure IV-55 and Figure IV-60 [101].

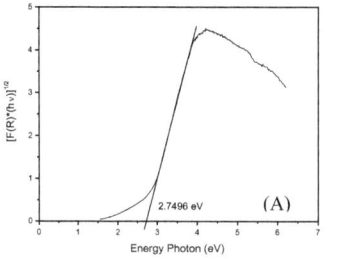 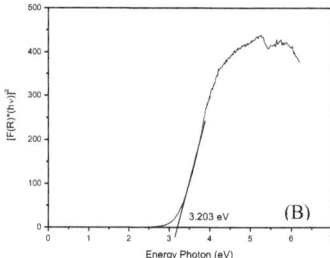

Figure IV-51. (A) $In_2Se_3$ indirect band gap determination and (B) $In_2Se_3$ direct band gap determination.

In Figure IV-50, the $In_2Se_3$ compound presents a higher absorption of visible light compared to the $TiO_2$ semiconductor because of the red shift shoulder which could drive to a possible absorption application of visible light. Therefore was considered the performance of $NO_x$ or acetaldehyde degradation experiments. Similar to the ones presented in the case of the S-doped $TiO_2$ material application (Figure IV-38 and IV-39). But the indium selenide compounds didn't show any degradation, neither of $NO_x$ nor acetaldehyde compounds, under UV-A light or visible light irradiation.

## Results and Discussion

To compare the material results after the heating procedure, SEM images were taken to find out, if any structural changes could be observed. In Figure IV-52 a more defined structural particle size of around 10 to 30 nm is noted. Which are bigger particles, than the ones obtained before the heating treatment, with particle sizes from 1.8 nm calculated after the XRD pattern (Figure IV.42) to the 10 nm found in Figure IV.44. After those comparisons, one argument for the lack of photoactivity of the $In_2Se_3$ compound could be expected due to the high amount of carbon surrounding the particles before the heating treatment. But even after removing the carbon from the $In_2Se_3$ surface, as can be seen in Figure IV.52 through the heating treatment, no activity was shown.

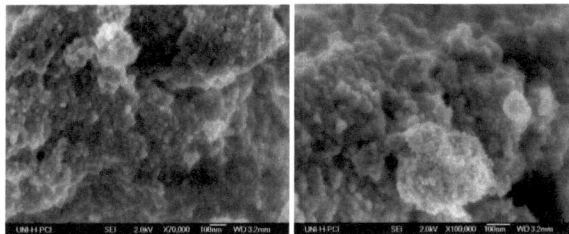

Figure IV-52. $In_2Se_3$ powder scan microscopy.

Since the synthesized indium selenide materials were not stable in water as tested for a DCA degradation experiment in aqueous phase, where no DCA decomposition took place after the TOC measurements, was necessary to explain the statement for this possible instability. The reason after the literature, is that the $In_2Se_3$ or $In_6Se_7$ material was decomposed in acidic solution forming $H_2Se$ [92]. It is shown, that photoreduction (photocorrosion) of Se in acid solution takes place as presented in Eq. IV.21.

$$Se + 2H^+ + 2e^- \longrightarrow H_2Se \qquad (IV.21)$$

In order to define experimentally a possible reason for these results, where no photocatalytic reaction in the aqueous or gas phase concerning the $In_2Se_3$ material occurred, two electrodes were prepared by a heat treatment deposition of $In_2Se_3$ on the conductive fluorine-tin oxide (FTO) glass at 500°C for one hour. The areas of the two working electrodes were 0.25 cm² and 1 cm².

## Results and Discussion

The thin film semiconductor electrodes were prepared on FTO glass type UTCO from AGC Fabritech Co Ltd. Japan with a 12 Ohm/cm² resistance and a thickness of 1.1 mm. The $In_2Se_3$ powder was suspended in acetone and the piece of FTO glass was immersed in the suspension. After the acetone was evaporated and the powder was fixed on the glass surface, the electrical contact was formed on the uncoated area of the substrate using a silver conductive paint and a copper wire. This contact area was later covered with nonconducting epoxy resin to isolate it from the electrolyte solution. A glass rod was placed over the copper wire for better handling. An elemental mapping analysis SEM was realized to confirm the homogeneity of the material over the glass surface (Figure IV-53). From the optical over view (Figure IV-53A) could be considered an heterogonous surface formed with agglomerated particles but through the elemental mapping analysis can be observed a very homogeneous distribution of the indium (Figure IV-53B) and the selenium (Figure IV-53C) elements over the whole glass electrode.

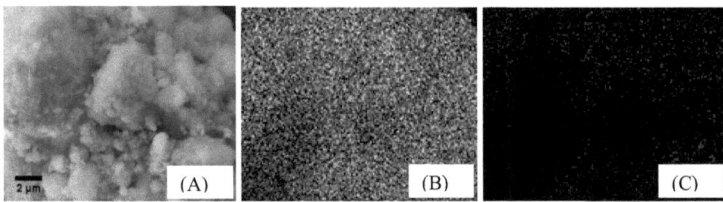

Figure IV-53. Elemental electron microscopy analysis of $In_2Se_3$ particles(A), corresponding indium mapping (B) and corresponding selenium mapping (C).

To find out the proportion of the elemental amount of the indium selenide compound on the piece of glass, was realized an EDXS analysis after the thermal treatment (500° C in the oven for one hour) (Figure IV-54). Here is observed a high difference in the proportion between the indium and selenium relation. Because of this unexpected relation, the peaks where quantified to have more information about the remained elements.

# Results and Discussion

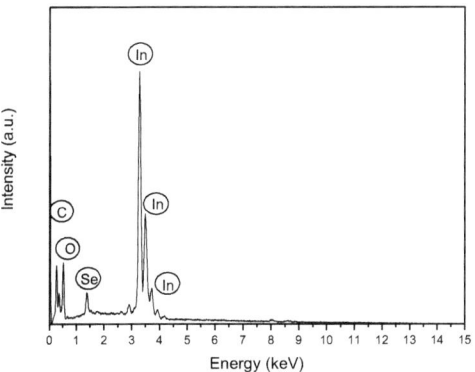

Figure IV-54. In$_2$Se$_3$ EDXS after 500°C calcination.

Table IV-14 shows a significant lower relation of selenium compared to the indium element amount, as suspected from the EDXS analysis in Figure IV-54. This selenium decrement is due the treatment in the oven at 500°C for one hour. This treatment represents a high lost of selenium, which should be avoided by a shorter period of heating in the oven. The expected relation of indium and selenium is usually a fifty-fifty relation of each of these elements.

Table IV-14. EDXS In$_2$Se$_3$ results.

| Element | expected weight (%) | obtained weight (%) |
|---|---|---|
| Se | 50.77 | 5.15 |
| In | 49.23 | 94.85 |
| Total | 100 | 100 |

## IV.5.3. Mott-Schottky study of the In$_2$Se$_3$ electrode

Even though knowing the low relation of In$_2$Se$_3$ on the electrode surface, resulted interesting to find out, if under illumination at the semiconductor-electrolyte interphase could be generated a charge transfer as photocurrent and photovoltage. The electrochemical setup consisted of three electrodes: the working electrode (semiconductor thin film), a platinum wire used as counter electrode and a saturated Ag/AgCl as reference (Figure III-7). The experiment was performed in aqueous 0.1 M KCl solution at pH 7. The potential was systematically varied between -1.00 and +1.00 V with the frequency range modulated between 100 Hz to 1 kHz. This maximum frequency value was considered because after the literature at higher frequencies like 10 kHz the systems should not be stable anymore showing complex behaviors like in the case of $TiO_2$ thin film electrodes [102].

The flatband potentials and the donor densities of $In_2Se_3$ thin films at the semiconductor/ electrolyte junction were obtained from Mott-Schottky plots (Figure IV-55 and Figure IV-60) (taken in the dark).
The semiconductor electrolyte interphase can be viewed as electrically equivalent to the two capacitances in series, (a) the space charge capacitance ($C_{SC}$) inside the semiconductor and (b) the Helmholtz layer capacitance ($C_H$) in the electrolyte region near the interphase. For a moderately concentrated electrolyte generally $C_{SC} \ll C_H$, and the total potential drop can be taken across $C_{SC}$ only. In terms of experimental voltages, the Mott-Schottky equation for the semiconductor-electrolyte interface is written as:

$$\frac{1}{C^2} = \frac{2}{\varepsilon_{InSe} \cdot \varepsilon_0 \cdot e_0 \cdot N_D} \cdot \left( E - E_{FB} - \frac{kT}{e_0} \right) \quad (IV.22)$$

$$\frac{1}{C^2}\left[\frac{1}{F^2 cm^4}\right] = \frac{2}{\varepsilon_{InSe} \cdot \varepsilon_0 \left[\frac{F}{m}\right] \cdot e_0[C] \cdot N_D[cm^3]} \cdot \left( E\left[\frac{J}{C}\right] - E_{FB}\left[\frac{J}{C}\right] - \frac{kT}{e_0}\left[\frac{J}{C}\right] \right) \quad (IV.23)$$

$$\frac{1}{C^2}\left[\frac{1}{F^2 cm^4}\right] = \frac{2}{4.9_{InSe} \cdot 8.854 \cdot 10^{-14} \left[\frac{F}{cm}\right] \cdot 1.60219 \cdot 10^{-19}[C] \cdot N_D[cm^3]} \cdot \left( E\left[\frac{J}{C}\right] - E_{FB}\left[\frac{J}{C}\right] - \frac{kT}{e_0}\left[\frac{J}{C}\right] \right) \quad (IV.24)$$

$$N_D = \frac{2}{\varepsilon_{InSe} \cdot \varepsilon_0 \cdot e_0 \cdot m_{slope}} \quad (IV.25)$$

$$E = E_{FB} + \frac{kT}{e_0} \quad (IV.26)$$

With:
$\varepsilon_0$ being the permittivity of free space, $\varepsilon$ $In_2Se_3$ the permittivity of the semiconductor electrode, $e_0$ the elementary charge, $N_D$ the donor density, E the applied potential, $E_{FB}$ the flatband potential, k the Boltzmann's Constant, T the temperature of operation, and C the space charge capacitance

Where $E$ is the applied voltage across the junction and $E_{FB}$ is the flat band potential or the electrode potential for which no potential drop occurs across the space charge region. A plot between $\dfrac{1}{C_{SC}^2}$ and $V$, commonly known as the Mott-Schottky plot, would be a straight line whose intercept on the $E$ axis yields $E_{FB}$, which is an important parameter for the solar cell fabrication. The donor concentration $N_D$ is calculated from the slope (7.53* $10^{13}$ F$^{-2}$cm$^{-4}$V$^{-1}$) of the curve. The dielectric constant $\varepsilon_s$ of the In$_2$Se$_3$ is taken as 4.9 [98]. Measurement of the flat band potential fixes the position of $E_F$ with respect to SCE.

From the Mott-Schottky diagram (Figure IV-55) related to the (0.25 cm²) In$_2$Se$_3$ electrode the donor density value ($N_D$) calculated was 6.11* $10^{18}$ cm³ and a flat band potential ($E_{FB}$) of -0.28 V was obtained. By comparing the donor density value of 1* $10^{18}$ cm³ from the literature [102] of a 1cm² TiO$_2$ thin film electrode at the same experimental conditions with this similar magnitude order value could further applications for the In$_2$Se$_3$ electrode be expected.

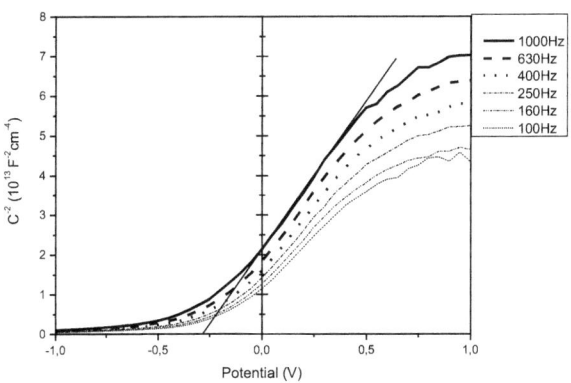

Figure IV-55. Mott-Schottky diagram from the 0.25 cm² In$_2$Se$_3$ electrode. Experiment performed in aqueous 0.1 M KCl solution at pH 7. With frequency range modulated between 100 Hz and 1 kHz.

# Results and Discussion

## IV.5.4. Current and photocurrent measurements of In$_2$Se$_3$ electrode

The current-voltage characteristics of the as deposited In$_2$Se$_3$ electrode (0.25 cm$^2$) in contact with the sodium selenate Na$_2$SeO$_4$ (1 mM) electrolyte or with Na$_2$S$_2$O$_3$ (0.1 M) electrolyte were studied in dark as well as under illuminated conditions (80 mW/cm$^2$) by using a 320 nm cut-off filter as shown in Figure IV-56 and Figure IV-57 respectively.

In the cathodic voltage region, current increases slowly whereas the anodic current initially increases slowly and then rapidly without any sign of saturation both in the dark as well as under illuminated conditions. A steep increase in the anodic current implies that there is a high density of surface states [103,104]. In the cathodic region as voltage becomes more negative, barrier height decreases and electrons from the conduction band are transferred through the space charge region into the electrolyte [101]. The anodic dark current is due to transfer of holes from the valence band to the electrolyte via surface states. At a smaller anodic potential, the holes are partly trapped in the surface states and then through inelastic processes are transferred to the electrolyte. As the potential increases, the traps are emptied and the current rapidly increases. When the junction is illuminated, electron hole pairs are generated in the valence band. For n-type material, the enhanced photocurrent/photovoltage is due to the minority carrier holes. Apart from the overlap of valence band edge and redox level, the hole transfer is also governed by the surface states and traps. The charge-transfer process remains the same as discussed in the case of the dark conditions. Thus an increase in current under illumination is observed.

A quantum yield ($\Phi$) of 8.96*10$^{-7}$ at a fixed electrical potential of 0.5 V was calculated (Eq. IV.27), from the experiment in Figure IV-56. It was realized with a 0.25 cm$^2$ In$_2$Se$_3$ electrode immersed in 1 mM Na$_2$SeO$_4$ electrolyte measuring the photocurrent density against the potential in the dark. This experiment was performed in the set up presented in Figure III-6.

$$\Phi = \frac{\frac{d[e]}{dt} \cdot A}{\frac{d[Photons]}{dt} \cdot A} \tag{IV.27}$$

$$\Phi = \frac{\frac{d[electrons]}{s} \cdot A}{\frac{d[Photons]}{s} \cdot A} \tag{IV.28}$$

## Results and Discussion

$$\Phi = \frac{i_e \left( \dfrac{\left(\dfrac{J}{s \cdot V \cdot cm^2}\right)}{1.60219 \cdot 10^{-19}(J)} \right) \cdot A(cm^2)}{i_{ph} \left( \dfrac{\left(\dfrac{J}{s \cdot cm^2}\right)}{\dfrac{h \cdot c}{\lambda}(J)} \right) \cdot A(cm^2)} \quad (IV.29)$$

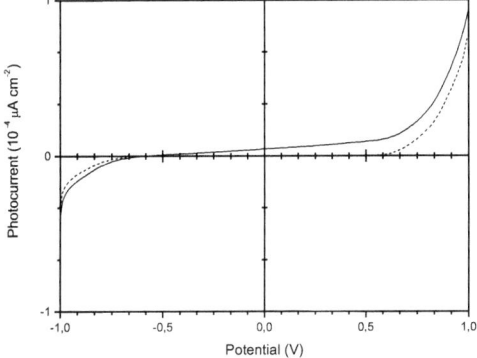

Figure IV-56. *J-V* characteristic in the dark (broken curve) and under illumination (full curve) of an 0.25 cm² In₂Se₃ electrode in 1 mM Na₂SeO₄ electrolyte. With an illumination intensity of 80 mW/cm².

In this case a quantum yield of $1.64 \times 10^{-6}$ (Eq. IV.27) was calculated at a fixed electrical potential of 0.5 V with a 0.25 cm² In₂Se₃ electrode. The difference in comparison to the last experiment in FigureIV-56 was the electrolyte used. It was immersed in a Na₂S₂O₃ 0.1 M electrolyte (Figure IV-57). This quantum yield value was obtained from measuring the photocurrent density against the potential in the dark.

# Results and Discussion

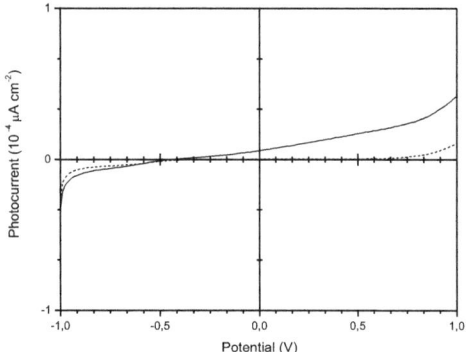

Figure IV-57. J-V characteristic in the dark (broken curve) and under illumination (full curve) of an 0.25 cm² $In_2Se_3$ electrode in $Na_2S_2O_3$ 0.1M electrolyte. The illumination intensity was 80 mW/cm².

A system which exhibits irreversible electrode dynamics will produce a cyclic voltammogram with different characteristics to those produced by a reversible system. The onset of the reaction will not occur at potentials immediately beyond the equilibrium potential. An appreciable overpotential will be required to induce the reaction as it is not kinetically favorable. As the overpotential is increased, the homogeneous rate constant increases and the rate of reaction will rise. The position of the peak maximum reflects the balance between the increasing heterogeneous rate equation and the decrease in the surface concentration of the reactant. The current is controlled by the electrode kinetics up to the peak maximum. After the peak maximum, the current simply reflects the rate at which the reactant diffuses though the solution to the surface of the electrode. Irreversible systems are not kept at equilibrium and the back reaction is negligible as occurred for example in the cyclic voltammograms in Figures IV-58 and IV-59. Reversible electrode behavior is the limiting case when the electrode kinetics are fast relative to the mass transport conditions. Conversely, irreversible electrode behavior is the limiting case when the electrode kinetics are slow relative to the mass transport conditions. Intermediate cases exist where the surface concentrations of the reduced and oxidized species depend on both the electron transfer rates (forward and reverse) and the rate of mass transport. In Figure IV-58 and Figure IV-59 are presented the voltammograms obtained of the $In_2Se_3$ electrode (0.25 cm²) immersed in $Na_2SeO_4$ 1 mM and sodium thiosulfate $Na_2S_2O_3$ 0.1 M respectively. It can be observed from both Figures that with a lower concentration of $Na_2SeO_4$

## Results and Discussion

electrolyte can be obtained a more stable and clearer voltammogram than with a higher concentration of the $Na_2S_2O_3$ electrolyte. In the reduction process (Figure IV-58) could be observed two defined couple of peaks which didn´t appear in the oxidation process, what concludes an irreversible system. In Figure IV-59 could not be notice the presence of any peak for the reduction side process, but one decreasing peak through the time in the oxidation reaction side. This defines the irreversibility of the whole redox process and therefore can be considered the $In_2Se_3$ as a instable material for an electrode application. It is clear, that each semiconductor material presents a better affinity for a specific electrolyte due to their particular redox properties. In this cases from Figures IV-58 and IV-59 with a small $In_2Se_3$ electrode (0.25 cm²) compared to a bigger one (1 cm²) (Figures IV-62 and IV-63), voltammograms showed that even using an electrolyte with a 100 times more dilluted electrolyte, but within the presence of one of the elements with which was built the electrode, two peaks were obtained showing a higher electron affinity. While with a concentrated sulfur electrolyte instead the presence of selenium, just one peak appeared in the compared reduction side process.

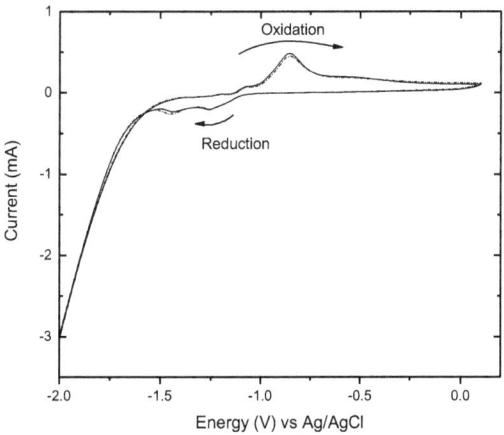

Figure IV-58. Cyclic voltammetry of a 0.25 cm² $In_2Se_3$ electrode immersed in 1 mM $Na_2SeO_4$ electrolyte (in the dark).

# Results and Discussion

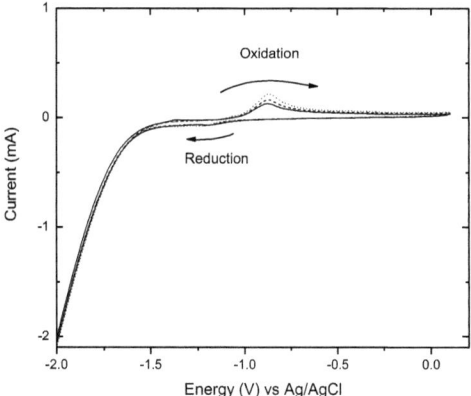

Figure IV-59. Cyclic voltammetry of a 0.25 cm² In₂Se₃ electrode immersed in Na₂S₂O₃ 0.1 M electrolyte (in the dark).

Supossing a possible increment of the $N_D$ value by increasing the electrode area was considered to repeat the same measurements, but in this case over an electrode with a surface of 1 cm². The intention was to find out if a lower concentration of the In₂Se₃ powder on the 0.25 cm² electrode was responsible for the low $N_D$ value and compare if any other parameter changed.

From the Mott-Schottky diagram (Figure IV-60) with the 1 cm² In₂Se₃ electrode was obtained a donor density value ($N_D$) of 4.48* 10¹⁷ cm³ and the flat band potential ($E_{FB}$) was -0.557 V. But as mention before from the smaller In₂Se₃ electrode (0.25 cm²) the obtained $N_D$ was 6.11*10¹⁸ cm³ and the $E_{FB}$ was -0.28 V (Figure IV-IV-55). After the experimental evaluation of both electrodes, was found that even though the electrode area was bigger the $N_D$ value resulted even lower in relation to the system with a smaller electrode area.

## Results and Discussion

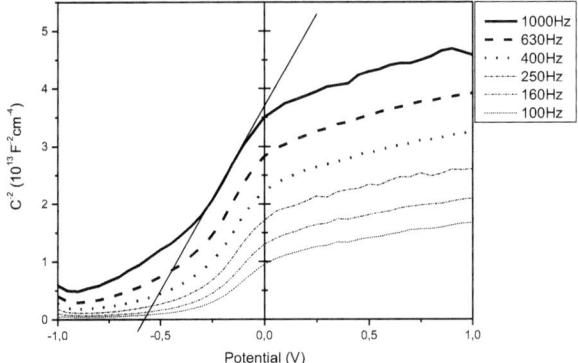

Figure IV-60. Mott-Schottky diagram of a 1 cm² In₂Se₃ electrode. Experiment performed in a 0.1 M KCl solution at pH 7 with a frequency range modulated between 100 Hz and 1 kHz.

$$\frac{1}{C^2} = \frac{2}{\varepsilon_{InSe} \cdot \varepsilon_0 \cdot e_0 \cdot N_D} \cdot \left( E - E_{FB} - \frac{kT}{e_0} \right) \quad \text{(IV.30)}$$

$$N_D = \frac{2}{\varepsilon_{InSe} \cdot \varepsilon_0 \cdot e_0 \cdot m_{slope}} \quad \text{(IV.31)}$$

$$E = E_{FB} + \frac{kT}{e_0} \quad \text{(IV.32)}$$

In the case of the 1 cm² In₂Se₃ electrode immersed in 1 mM sodium selenate electrolyte (Figure IV-61) a higher photocurrent was obtained in comparison to the smaller electrode (Figure IV-56). The quantum yield at a fixed electrical potential of 0.5 V was 1.52*10⁻⁶ (Eq. IV.27). It corresponds to the experiment with 1 cm² In₂Se₃ electrode immersed in 1 mM Na₂SeO₄ electrolyte for the photocurrent density against the potential in the dark.

## Results and Discussion

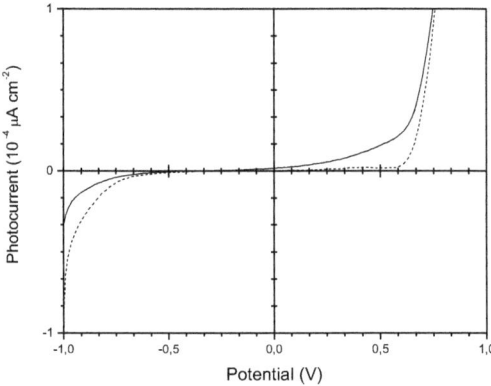

Figure IV-61. *J-V* characteristic of an $In_2Se_3$ (1 cm$^2$) electrode in the dark (broken curve) and under an illumination intensity of 80 mW/cm$^2$ (full curve) in $Na_2SeO_4$ 1 mM electrolyte.

By comparing the 0.25 cm$^2$ and the 1 cm$^2$ electrodes, the voltammogram with the higher surface present's higher couple of peaks for both electrolytes what makes sense, because with a higher surface is possible to have higher contact with the electrolyte and therefore get a higher electron exchange. In the case of the 1 cm$^2$ electrode the peak also increases with a higher concentration of electrolyte, something that is not possible to be appreciated by testing the smaller electrode.

The results obtained from the 1 cm$^2$ $In_2Se_3$ electrode in Figure IV-62 and Figure IV-63 of the cyclic voltammetry were analogous to the ones obtained from the smaller surface area electrode with 0.25 cm$^2$ (Figures IV-58 and IV-59). In both cases the material was not stable what obviously is not related to the electrode size or in other words to the concentration of the $In_2Se_3$ powder film in contact with the electrolyte. The stability depends just on the $In_2Se_3$ material which decomposes easily in the oxidation process.

Results and Discussion

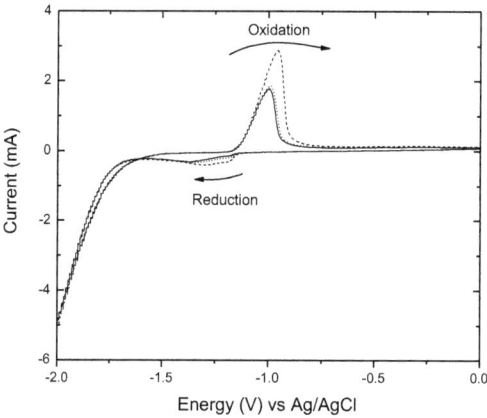

Figure IV-62. Cyclic voltammetry of 1 cm² In$_2$Se$_3$ electrode in Na$_2$SeO$_4$ 1 mM electrolyte.

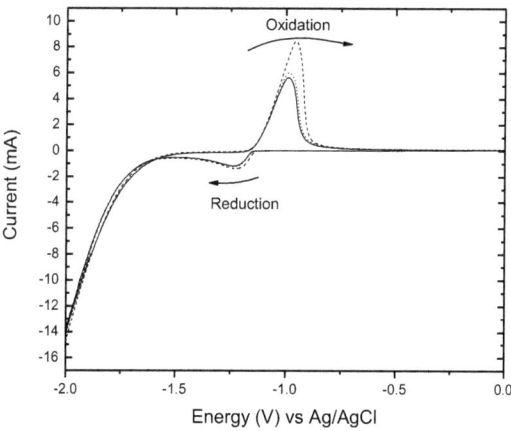

Figure IV-63. Cyclic voltammetry of 1 cm² In$_2$Se$_3$ electrode in Na$_2$S$_2$O$_3$ 0.1 M electrolyte.

## IV.5.5. Conclusions

$In_2Se_3$ was successfully synthesized. Through the characterization was possible to confirm its presence and find out important properties for further applications of this material. By changes in the synthesis procedure was feasible to obtain different properties related to its structure. The annealing temperature for the synthesis of indium selenide compounds was a very important variable to define, if it would be a p- or n-type semiconductor, what results of high consideration to get appropriate and convenient advantages.

With this synthesis was possible to obtain $In_2Se_3$ or $In_6Se_7$ depending on the period of reaction, at relative lower temperatures (300 °C) and with a particle size of 1.8 nm in the case of the $In_2Se_3$.

By increasing the periods of reaction in the synthesis can be also modified the stoichiometry.

The band gap for the $In_2Se_3$ was estimated as 2.74 eV and the material was considered an indirect semiconductor after Kubelka-Munk calculations of the diffuse reflectance spectra.

Photocatalytic decomposition reactions were performed in aqueous phase but the material resulted not stable enough. In the gas phase was pretended to decompose acetaldehyde or $NO_X$ but no activity was observed. The gas phase reactions were executed with 50 % humidity what could have an inconvenient influence for the inactivation of this compound as could be observed the instability of the material in the voltammograms, where $In_2Se_3$ electrodes were immersed in aqueous electrolytes.

Because of the positive inclination slope in the Mott-Schottky diagram can be confirmed that the synthesized material $In_2Se_3$ corresponds to a n-type semiconductor [101].

## Results and Discussion

IV.6. Solid state synthesis of $\beta$-$Ga_2O_3$ by heat treatment and characterized as a film electrode or powder showing photocatalytic improvement decomposing acetaldehyde

Gallium oxide ($Ga_2O_3$) can adopt five different crystalline structures, which are, $\alpha$-, $\beta$-, $\gamma$-, $\delta$- and $\varepsilon$- phases. The $\beta$-$Ga_2O_3$ structure appears to be quite different from that of $\alpha$-$Ga_2O_3$ which has the $\alpha$-corundum structure [105]. The latter one has the oxygen ions in approximately hexagonal close-packed array with all the $Ga^{3+}$ ions octahedrally coordinated to $O^{2-}$ ions. Also in the $\alpha$ phase, octahedral share edges and faces which brings the metal ions very near to each other. In $\beta$-$Ga_2O_3$ no faces are shared between polyhedral and the shortest $Ga^{3+}$-$Ga^{3+}$ distance in 3.04 A. Now it is recognized that usually structures in which faces of polyhedral are shared are less stable than those in which edges are shared, which in turn are less stable than structures in which only corners are shared. Thus, one would expect the $\beta$ phase to be more stable than the $\alpha$ phase. Foster and Stumpf have shown that although the $\alpha$- $Ga_2O_3$ forms at lower temperatures than does $\beta$- $Ga_2O_3$ the $\alpha$ phase is metastable [106]. Among these phases, $\beta$-$Ga_2O_3$ is an important wide band-gap compound and its band-gap is 4.9 eV. It is well known that $\beta$-$Ga_2O_3$ semiconductors have been widely studied not only for its high stability and tenability in optical properties, but also for the applications in highly sensitive oxygen gas sensors, and transparent conductors in optoelectronics [107]. Where recently, thin film gas sensors have attracted special interests like semiconducting metal oxide sensors. Some of them have been commercialized such as an $SnO_2$ sensor for domestic gas leakage alarm, a solid electrolyte $ZrO_2$ sensor detecting oxygen concentration in an automobile exhaust pipe, and a burn sensor of a $TiO_2$ thick film. $Ga_2O_3$ thin film is considered one of the most favorite materials for high-temperature-stable gas sensor [108], and because of this stability would be interesting to test this metal oxide also as a photocatalyst in the gas phase. There are few investigations considering photochemical decomposition of damaging environmental compounds in aqueous phase with $\beta$-$Ga_2O_3$ [109] but none to the best knowledge of the authors considering the gas phase. However, only a limited number of photocatalysts have been reported to be photocatalytically active for the $H_2O$ splitting reaction. Recently the series of oxides and mixed oxides of P-block elements with $d^{10}$ electron configurations, such as the oxides containing Ga, Ge, In, Sn, and Sb are reported to be able to catalyze this reaction effectively [110].

## Results and Discussion

### IV.6.1. Synthesis of gallium acetate

It was necessary to self synthesize the gallium acetate compound, because the product could be just delivered after three months. The synthesis of gallium acetate was produced by adding 2.5 g $GaCl_3$ in a solution of 10-20 mL ether with 50 mL glacial acetic acid. The replacement of the third chloride atom through the acetate rest takes longer to be effected. The ether was distilled off in a nitrogen atmosphere and the solution was heated to the boiling point under reflux. The reaction took uninterrupted one week maintaining the temperature at the boiling point. The precipitated crystal powder were filtrated, washed with ether and finally dried under nitrogen stream. The triacetate is soluble in water with an acidic reaction but insoluble in alcohol or ether [111].

### IV.6.2. Synthesis and characterization of $\beta$-$Ga_2O_3$

The gallium triacetate was heated and dissolved in 5 mL dodecylamine to be injected dropwise into a 280° C heated hexadecylamine solution (50 mL) and the temperature was maintained for 24 hours. After cooling, 100 mL methanol were added producing a mixture where a black viscous solid-gel was isolated by centrifugation.

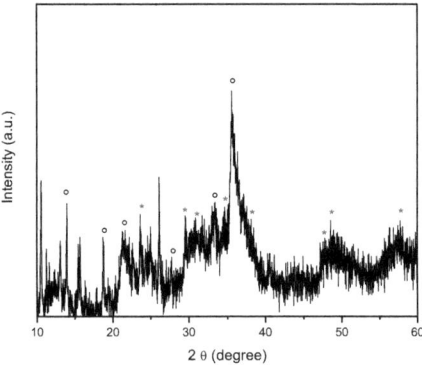

Figure IV-64. X-ray diffraction pattern of $Ga_2O_3$ sample (° peaks correspond to $\alpha$-$Ga_2O_3$ and * peaks correspond to $\beta$-$Ga_2O_3$ after WinXPOW database).

## Results and Discussion

The synthesized material under the reaction conditions showed a low crystalline structure (Figure IV-64). But due to the peaks can be suggested the formation of $Ga_2O_3$. To be more precise, after the literature was possible to identify the product as $\alpha$-GaO(OH) [112]. The low intensity peaks could be considered, as evidence of the starting conversion point into the $\beta$-$Ga_2O_3$ compound. From the Sherrer equation (IV.12) was obtained a particle size of 1.69 nm.

The obtained material was analyzed by SEM to find out, if any crystalline structure was present after the synthesis and confirm the XRD results of Figure IV-64. From the SEM observation was revealed the possible growth of bigger clusters and a high roughness on the surface. But it was not possible to see any define structure of the $Ga_2O_3$ compound (Figure IV-65) because the particles seem to be covered by an amorphous layer of carbon. This carbon remained from the added hexadecylamine during the synthesis procedure and therefore was not possible to have a higher resolution of the Figure IV-65.

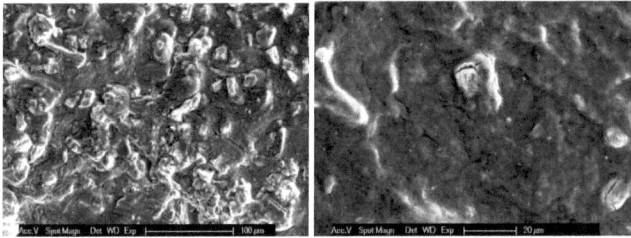

Figure IV-65. SEM $Ga_2O_3$.

The $Ga_2O_3$ compound was analyzed by EDXS (Figure IV-66), where some chlorine is to be observed after the synthesis. Gallium, oxygen and carbon were present. With this information the next step to be followed was to calcinate the $Ga_2O_3$ to purify it and obtain a higher crystalline structure.

## Results and Discussion

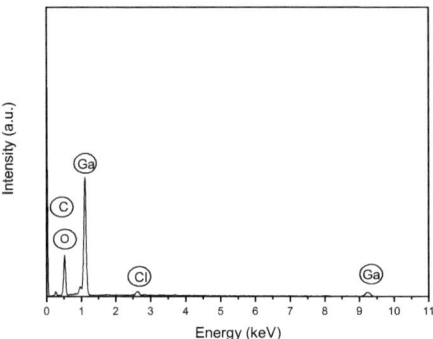

Figure IV-66. $Ga_2O_3$ EDXS analysis after synthesis.

The material was calcinated at 500° C for one hour in order to get the rest of chloride molecules released from the structure (Figure IV-67). The obtained structure presents defined peaks corresponding to a $\beta$-$Ga_2O_3$ [113] arrange with a particle size of 5.14 nm calculated by the Sherrer equation (IV.12). The resulting material appearance changed from a black viscose consistency to a gray powder.

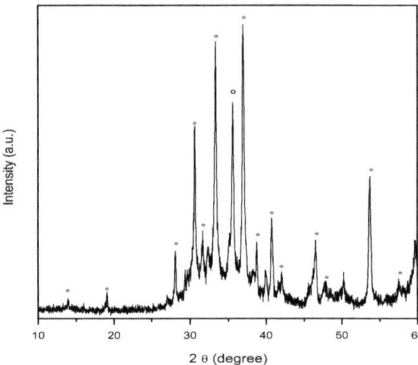

Figure IV-67. XRD of the synthesized $Ga_2O_3$ powder after one hour of calcination at 500°C sample (° peak correspond to $\alpha$- $Ga_2O_3$ and * peaks correspond to $\beta$-$Ga_2O_3$ after WinXPOW database).

## Results and Discussion

The β-Ga$_2$O$_3$ powder was calcinated at 500 °C for one hour and observed under SEM to compare any change due to the temperature increase treatment. Through the different XRD patterns before (Figure IV-64) and after the calcination (Figure IV-67), could be clearly observed an increase in the crystallinity. But this change was not possible to be observed in the scanning microscopy analysis (Figure IV-68).

Figure IV-68. β-Ga$_2$O$_3$ scan microscopy powder calcinated at 500° C for one hour.

In Figure IV-69 could be observed through the elemental analysis the total chlorine removal of the gallium oxide powder, which was present in the EDXS before calciantion (Figure IV-66). It could also be confirmed that the carbon element was not completely removed from the Ga$_2$O$_3$ compound making difficult the SEM study in Figure IV-68.

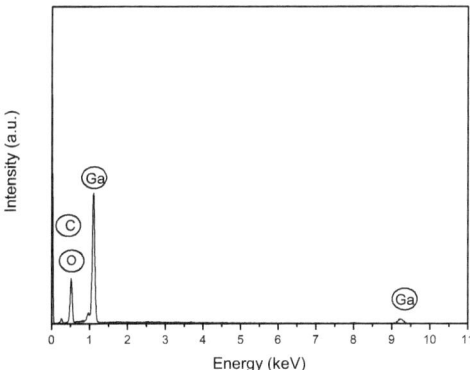

Figure IV-69. EDXS after one hour calcination at 500° C.

## Results and Discussion

In Figure IV-70 is presented the X-ray diffraction pattern of the gallium oxide powder calcinated again at 500° C but in this case for five hours in order to increase the crystallinity, which apparently didn't change drastically by comparing the XRD pattern in Figure IV-67. The particle size obtained (9.41 nm) in Figure IV-70 increased almost twice as the calculated from the pattern in Figure IV-67 by the Sherrer equation (IV.12).

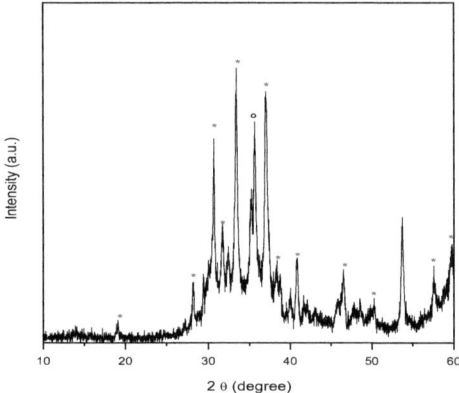

Figure IV-70. XRD of $Ga_2O_3$ calcinated at 500°C for 5 hours (° peak correspond to $\alpha$- $Ga_2O_3$ and * peaks correspond to $\beta$-$Ga_2O_3$ after WinXPOW database).

Although there was no apparent change from the X-ray diffraction patterns between the calcination of one or five hours at 500° C, it was interesting the morphology resulted in Figure IV-71 of the STEM gallium oxide particles calcinated for 5 hours at 500°C. Here in Figure IV-71 (A) is possible to observe the organized structure of the $\beta$-$Ga_2O_3$ as nanofibres which grow from some hundreds of nanometers to a length of 5 micrometers. Figures IV-71 (B) (C) and (D) present in detail the structures of a short and long arrangement to sustain the growing hypothesis according to aggregation of thin fibers as can be clearly follow in Figure IV-71 (D). Similar structures were not observed previously, just after this 500 °C calcination.

## Results and Discussion

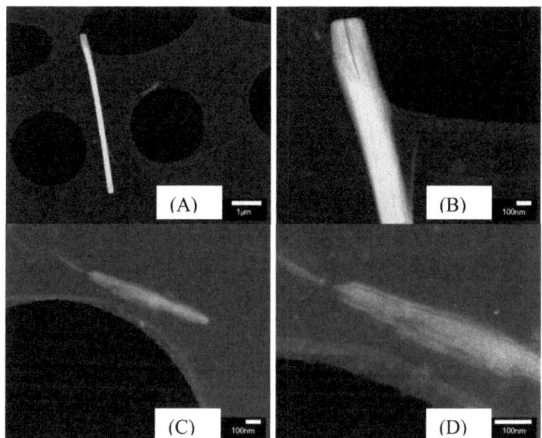

Figure IV-71. STEM nanofibres of gallium oxide calcinated at 500°C for 5 hours.

The gallium oxide nanofibres were calcinated again but now at 1000° C for 5 hours. The XRD pattern of $\beta$-$Ga_2O_3$ is shown in Figure IV-72. The strong and sharp diffraction peaks indicate a high degree of crystallization. All the peaks in the XRD pattern were indexed as $\beta$-$Ga_2O_3$ in the WinXPOW database and no other crystalline phases were detected.

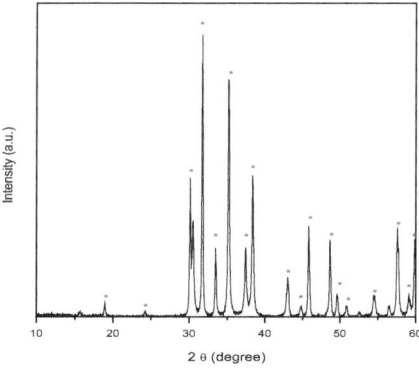

Figure IV-72. XRD pattern of $\beta$-$Ga_2O_3$ calcinated for 5h at 1000°C (* peaks correspond to $\beta$-$Ga_2O_3$ after WinXPOW database).

## Results and Discussion

The XRD results, revealed that nanofibres after calcinated at 1000° C were indeed $\beta$-$Ga_2O_3$ [114] and other form of crystalline gallium oxide (if any) were below the detection level. The particle size (14.81 nm) was obtained from the calculation of the Sherrer equation (IV.12).

In Figure IV-73 is presented a theoretical crystal structure of $\beta$-$Ga_2O_3$ defined by calculated translations vectors [115], where the primitive unit cell of this structure contains only half as many atoms as the conventional one, due to the oxygen vacancy molecules in the structure.

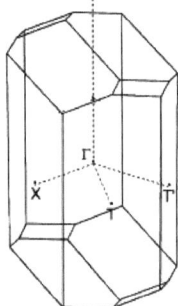

Figure IV-73. Brillouin zone structure of $\beta$-$Ga_2O_3$. The k points that overlap in the $\Gamma$ point of the supercell are marked with different letters.

TEM and HRTEM images of $\beta$-$Ga_2O_3$ in Figure IV-74 showed a change in the structure from fibers to nanoparticles within a size average of 100 nm and a well defined crystal structure after being calcinated at 1000 °C.

Results and Discussion

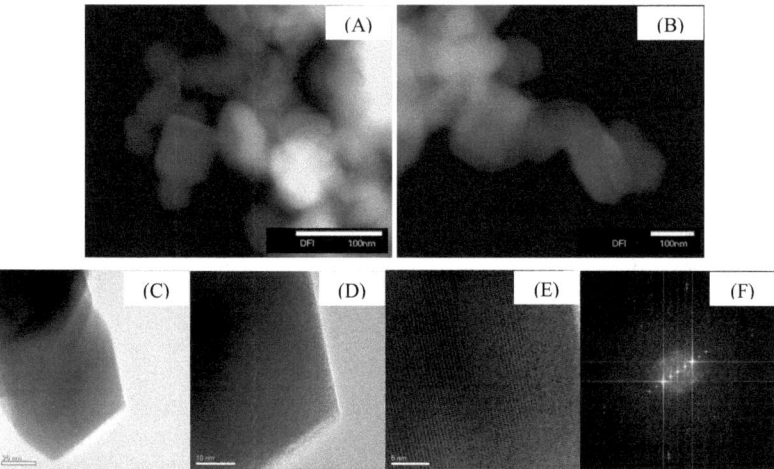

Figure IV-74. $\beta$-Ga$_2$O$_3$ TEM pictures are presented in (A) and (B) and HRTEM in (C),(D),(E) and in (F) is presented the optical diffraction with a grid distance of $d_{011}$ 2.495 Å.

By characterizing $\beta$-Ga$_2$O$_3$ (Figure IV-75) was determined and compared the reflectance diffraction as the Kubelka-Munk function with other semiconductors to obtain experimental values to reinforce the data of the literature and define the synthesized compound. The dark gray powder calcinated for 1 hour at 500°C presents an unusual reflectance diffraction showing a very high absorption but after being calcinated at 1000° C turned into a white powder becoming a perfect crystalline structure.

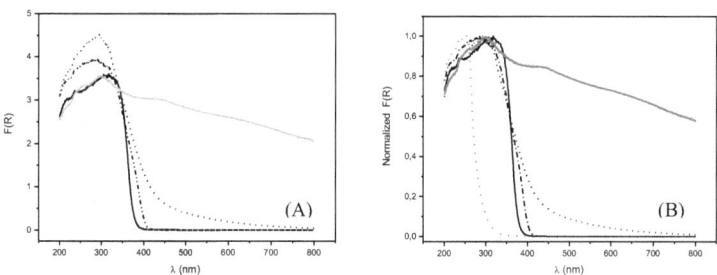

Figure IV-75. Reflectance function (A) and of normalized Reflectance function (B) of $\beta$-Ga$_2$O$_3$ calcinated at 1000° C (∘∘∘), anatase PC500 (—), rutile R15 (---), In$_2$Se$_3$ (•••) and Ga$_2$O$_3$ calcinated at 500° C (—) are compared.

## Results and Discussion

Direct band gap calculation for $\beta$-Ga$_2$O$_3$ after calcinated at 1000° C for 5 hours is presented in Figure IV-76 (A) and the indirect band gap calculation in Figure IV-77 (B). The direct band gap tangent calculation fits better to consider the $\beta$-Ga$_2$O$_3$ compound a direct semiconductor.

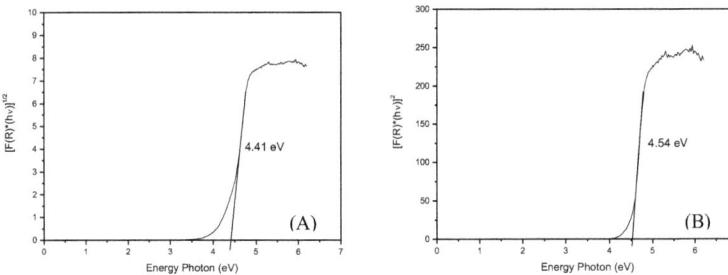

Figure IV-76. (A) $\beta$-Ga$_2$O$_3$ indirect band gap determination (B) $\beta$-Ga$_2$O$_3$ direct band gap determination.

To establish if this semiconductor corresponds to a p- or n-type should be considered the literature to determine this property. One suggestion came from the photoluminescence properties of $\beta$-Ga$_2$O$_3$ which have been extensively studied for several decades and acceptable model for blue has been put forward [116]. Liang *et al.*, had obtain two emission peaks at 380 nm and 525 nm where the blue band emission (466 nm) may originate from the recombination of an electron on a donor formed by oxygen vacancies (V$_O$) and a hole on an acceptor formed by gallium vacancies (V$_{Ga}$) or by gallium-oxygen vacancy pairs (V$_O$,V$_{Ga}$) [117]. With calcinating at high temperature as in the present work, oxygen vacancies can easily be form. It is expected that the formation of O vacancies (V$_O$) or vacancy pairs (V$_O$, V$_{Ga}$) were present in this synthesis of $\beta$-Ga$_2$O$_3$ nanoparticles at high temperature, which supports the origin of blue emission forming a p-type semiconductor. The most appropiate measurement to find out, if this material corresponds in deed to a p-type is to prepare a film and carry out Hall measurements. This was not possible due to the lack of time.

## Results and Discussion

### IV.6.3. Photocatalytic decomposition of acetaldehyde on $\beta$-Ga$_2$O$_3$ nanoparticles under UV-A light

An acetaldehyde degradation experiment (Figure IV-77) was performed using $\beta$-Ga$_2$O$_3$ nanoparticles (calcinated for about 5 hours at 1000°C) as photocatalyst under 1mW/cm² UV-A light irradiation. The same experiment was tested under visible light but there was not found any photoactivity. Because of the wide band gap of $\beta$-Ga$_2$O$_3$ should be UV-C light applied for a better performance of this photocatalyst according to its absorption range but with the intention of comparing later on, the results of the other developed photocatalysts was applied this lower intensity. Experiments with the Ga$_2$O$_3$ material without calcination treatment were also studied but any acetaldehyde degradation reaction was observed.

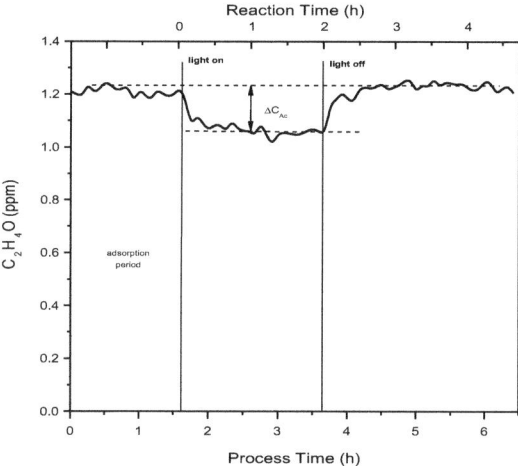

Figure IV-77. Acetaldehyde degradation with 1g pressed $\beta$-Ga$_2$O$_3$ calcinated (for 5h at 1000°C) photocatalyst under 1mW/cm² UV-A light.

In Table IV-15 is presented the calculated result of the photonic efficiency after decomposing acetaldehyde in gas phase using $\beta$-Ga$_2$O$_3$ as photocatalyst under UV-A light (360 nm). Before the

## Results and Discussion

photocatalyst calcination was no photoactivity shown, even after the 500°C. But after 1000°C calcination was possible to observe a degradation of the model compound.

Table IV-15. Photonic efficiency result of acetaldehyde degradation under UV light with a calcinated $\beta$-$Ga_2O_3$ photocatalyst.

| Photocatalyst | (UV-A light) | |
|---|---|---|
| | glass reactor | |
| | Rate [mol/s] | Intensity [mol/s] |
| $\beta$-$Ga_2O_3$ (1.0g) | 1.16*10⁻¹⁰ | 1.19*10⁻⁷ |
| % Ph. Efficiency | 0.097 | |

### IV.6.4. Preparation of $Ga_2O_3$ electrode

After characterizing the initial fresh synthesized gallium oxide powder was decided to make a film, where a conductive fluorine-tin oxide (FTO) glass type UTCO from AGC Fabritech Co Ltd. Japan with a 12 Ohm/$cm^2$ resistance and a thickness of 1.1 mm, was coated and calcinated for 1 hour at 500° C in order to utilize the semiconductor thin film as electrodes, they were prepared on and the piece of FTO glass which was immersed in a suspension of $Ga_2O_3$ in toluene. After the toluene was evaporated and the powder was fixed on the glass surface the electrical contact was formed on the uncoated area of the substrate using a silver conductive paint and a copper wire. This contact area was later covered with nonconducting epoxy resin to isolate it from the electrolyte solution. A glass rod was placed over the copper wire for better handling.

### IV.6.5. Characterization of the $Ga_2O_3$ electrode

An elemental mapping analysis SEM was realized to confirm the homogeneity of the material over the glass surface (Figure IV-78). With an upper view of the coated surface is observed a well dispersion of the semiconductor material over the film. And with a side view of the coated glass can be appreciated the different elemental mapping layers in (Figure IV-79).

Results and Discussion

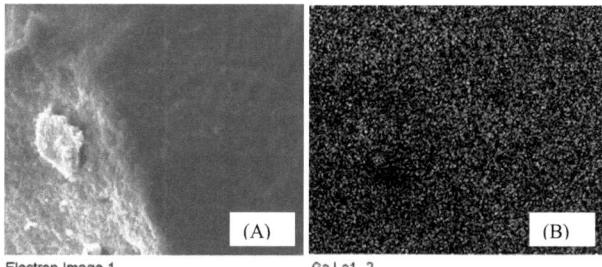

Figure IV-78. Elemental electron microscopy analysis of $Ga_2O_3$ film (upper view) (A) and corresponding gallium mapping (B).

Here (Figure IV-79) is possible to appreciate the composition of each layer of the film after 1 hour of calcination at 500°C (one side view).

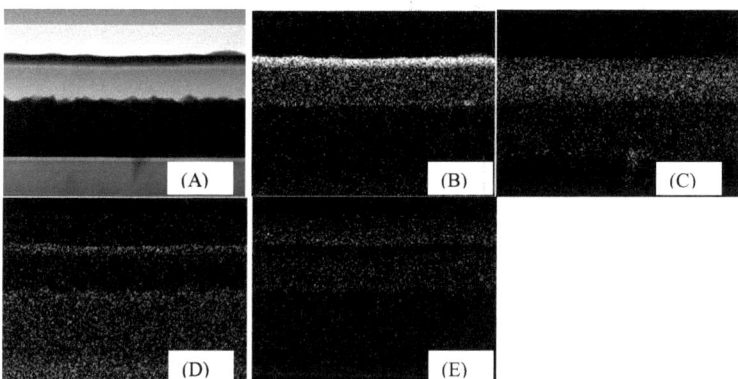

Figure IV-79. Elemental electron microscopy analysis of $Ga_2O_3$ film (side view) (A), gallium mapping (B), chlorine mapping (C), oxygen mapping (D) and carbon mapping (E).

The EDX spectra detection of the film in Figure IV-80 confirms the elements detected in the powder but on the film the elements are stratified on different layers, due to the gradient of temperature established between the layers, what results of high importance for the reaction but also as a parameter of isolation and resistance presented by the $Ga_2O_3$ compound, especially considering the thickness of the gallium oxide layer in comparison to the others, where this one appear to be the thinner one.

135

Results and Discussion

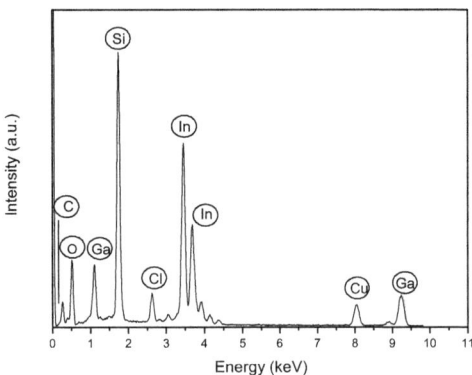

Figure IV-80. EDXS of $Ga_2O_3$ film calcinated for one hour at 500°C.

## IV.6.6. Mott-Schottky study of the $Ga_2O_3$ electrode

From the Mott-Schottky diagram (Figure IV-81) of the $Ga_2O_3$ electrode (0.25 cm²) was obtained the donor density value ($N_D$) as (-1.55* 10¹⁶ cm³) and the flat band potential ($E_{FB}$) as (-0.887 V).

The donor concentration $N_D$ is calculated from the slope (-8.61* 10¹⁵ F⁻²cm⁻⁴V⁻¹) of the curve. The dielectric constant $\varepsilon_s$ of the $Ga_2O_3$ is taken as 10.4 [118,119], although $Ga_2O_3$ possesses a dielectric constant of 10.2-14.2. Measurement of the flatband potential fixes the position of $E_F$ with respect to SCE.

Gallium oxide exhibits p-type semiconductor property at the temperature of 500° C [105], because of the oxygen vacancy in the crystal, which then transforms into a $\beta$-$Ga_2O_3$ after calcinated at 1000°C condition which can be more favorable for the creation of more vacancies in the structure.

Results and Discussion

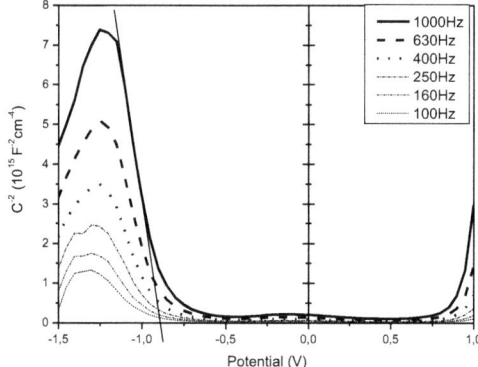

Figure IV-81. Mott-Schottky diagram of a 0.25 cm² Ga₂O₃ electrode. immersed in a 0.1 M KCl solution at pH 7 with a frequency range modulated between 100 Hz and 1 kHz.

$$\frac{1}{C^2} = \frac{2}{\varepsilon_{InSe} \cdot \varepsilon_0 \cdot e_0 \cdot N_D} \cdot \left( E - E_{FB} - \frac{kT}{e_0} \right) \quad (IV.33)$$

$$N_D = \frac{2}{\varepsilon_{GaSe} \cdot \varepsilon_0 \cdot e_0 \cdot m_{slope}} \quad (IV.34)$$

$$E = E_{FB} + \frac{kT}{e_0} \quad (IV.35)$$

## IV.6.7. Current and photocurrent measurements of Ga₂O₃ electrode

From the (0.25 cm²) Ga₂O₃ electrode immersed in a sodium thiosulfate electrolyte (Na₂S₂O₃ 0.1 M) under 80 mW/cm² illumination increases the previously established potential in the dark what can be measured as the generated photocurrent in the system (Figure IV-82). It can be observed that by increasing the potential the photocurrent increases considerably more than the potential obtained from the dark reaction, what makes interesting this material for solar applications. Where a higher potential energy can be obtained due to the efficient absorption of photons on the electrode surface.

## Results and Discussion

The quantum yield ($6.52*10^{-7}$) obtained (Eq. IV.27) from the experiment with an 0.25 cm² $Ga_2O_3$ electrode immersed in 0.1 M $Na_2S_2O_3$ electrolyte in Figure IV-82 was calculated for the photocurrent density against the potential at a fixed electrical potential of 0.5 V.

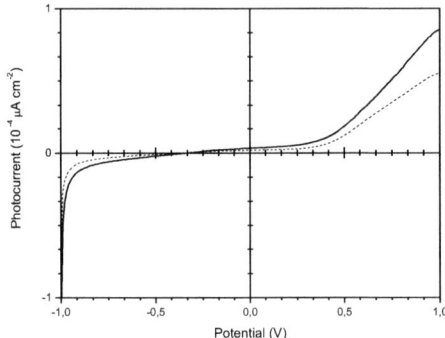

Figure IV-82. *J-V* characteristic of a 0.25 cm² $Ga_2O_3$ electrode in the dark (broken curve) and under illumination (80 mW/cm²) (full curve) immersed in a 0.1 M $Na_2S_2O_3$ electrolyte.

By comparing the results between the photocurrent and the potential in the dark of a 0.25 cm² $Ga_2O_3$ electrode immersed now in a sodium selenate electrolyte (1 mM $Na_2SeO_4$) under 80 mW/cm² irradiation, can be observed a very small potential increment under illumination (Figure IV-83). One reason for a low photocurrent and potential in the dark corresponds to the lower electrolyte concentration. The use of a higher concentration of this electrolyte is not convenient because could result very toxic and also damage the environment in case of future applications.

The quantum yield ($3.52*10^{-7}$) obtained (Eq. IV.27) from the experiment with an 0.25 cm² $Ga_2O_3$ electrode immersed in 1 mM $Na_2SeO_4$ electrolyte in Figure IV-83 was calculated for the photocurrent density against the potential at a fixed electrical potential of 0.5 V.

## Results and Discussion

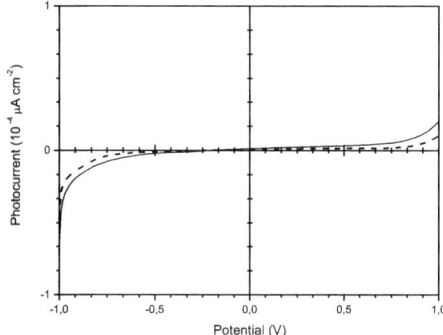

Figure IV-83. *J-V* characteristic of a 0.25 cm$^2$ Ga$_2$O$_3$ electrode in the dark (broken curve) and under illumination (80 mW/cm$^2$) (full curve) immersed in 1 mM Na$_2$SeO$_4$ electrolyte.

To establish the correspondent importance of the electrolyte in the system was carried out a cyclic voltammetry measurement with the Ga$_2$O$_3$ electrode.

Cyclic voltammograms were recorded with a computer controlled E-Chemie $\mu$-Autolab potentiostat.

This technique was used to study the electrode redox couple in a three electrode system. The potential of the working electrode is swept between two set potentials. The scan rate (the magnitude of the rate of change of the applied potential with time) is kept constant.

As the applied potential is increased, oxidation will occur and a positive current will flow (electron transfer from the species being oxidized to the electrode). Conversely, as the potential is decreased, reduction will occur and a negative current will flow (electron transfer from the electrode to the species being reduced).

In the case of the reversible system, the electron transfer rates at all potentials are significantly greater than the rate of mass transport, and therefore Nernstian equilibrium is always maintained at the electrode surface. When the rate of electron transfer is insufficient to maintain this surface equilibrium then the shape of the cyclic voltammogram changes a little. The most marked feature of a cyclic voltammogram of a totally irreversible system is the total absence of a reverse peak. However is important to consider, such a feature on its own does not necessarily imply an irreversible electron transfer process, but could be due to a fast following chemical reaction. In the case of the Ga$_2$O$_3$ electrode immersed in a 0.1 M Na$_2$S$_2$O$_3$ electrolyte (Figure IV-84) can be observed the oxidation process by the release of two electrons, but as can be clearly seen, the reduction process it does not involve any gain of electrons.

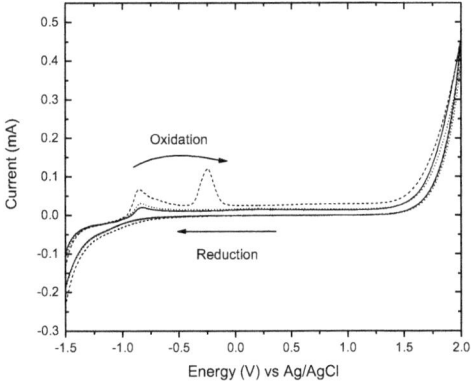

Figure IV-84. Cyclic voltammetry of a 0.25 cm² $Ga_2O_3$ electrode immersed in 0.1 M $Na_2S_2O_3$ electrolyte.

In the oxidation and reduction process of a 0.25 cm² electrode $Ga_2O_3$ immersed in a 1 mM $Na_2SeO_4$ electrolyte (Figure IV-85) did not appear any electron exchange at all.

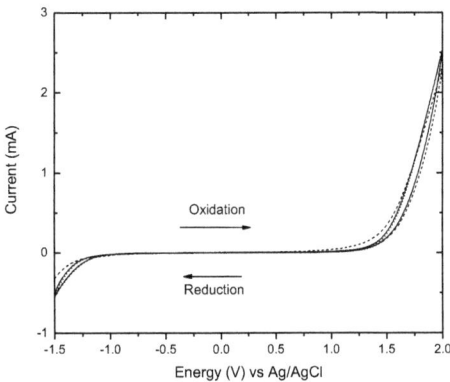

Figure IV-85. Cyclic voltammetry of a 0.25 cm² $Ga_2O_3$ electrode immersed in 1 mM $Na_2SeO_4$ electrolyte.

## IV.6.8. Conclusions

The band gaps obtained of the synthesized $\beta$-$Ga_2O_3$ compound were 4.41 eV and 4.54 eV for the indirect band gap and for the direct band gap respectively. Values which were consistent with the reported ones 4.2-4.9 eV depending upon the preparation conditions [115]. The band gap energy value of 4.4 eV was earlier determined for films and for $Ga_2O_3$ bulk [120]. The $\beta$-$Ga_2O_3$ compound was defined as a direct p-type semiconductor. After the best fit of the tangent calculations from the Kubelk-Munk diffusion reflectance spectrum was defined a direct band gap. And the Mott-Schotttky plot showed a negative inclination slope what corresponds to a p-type semiconductor.

With this synthesis was possible to obtain $\beta$-$Ga_2O_3$ nanoparticles with sizes of 1.7 nm. It resulted difficult to remove the organic substances used during the synthesis. Therefore was planned to remove the rest of the organic substances by increasing the temperature while simultaneously should increase the crystallinity of the compound what succeded after heating up to 1000 °C for five hours. With this obtained white $\beta$-$Ga_2O_3$ treated powder, unexpectedly, even though it's wide band gap was possible to decompose acetaldehyde under the irradiation of UV-A light.

In the literature also even the wide band gap $\beta$-$Ga_2O_3$ is found as one of the promising photocatalyst for over all water splitting of $H_2O$. The photocatalyst can easily improve its performance by the addition of small amount of metal ion like Ca, Sr, Ba, Cr, Ta and Zn, particularly the addition of Zn ion gives a remarkable effect to improve the photocatalytic activity [110].

Conclusions

## V. General conclusions

Nanoparticles or thin film photoelectrodes are needed in photoelectrocatalytic systems to apply a bias potential, either for photoelectrode characterization or to facilitate the photocatalytic reactions. Their electronic band structure (i.e. both the band gap energy and the positions of CB and VB) represent the key factor to determine whether or not a semiconductor material is suitable for a specific photocatalytic reaction.

It is known that the $U_{fb}$ for most semiconductors, such as n- and p-GaAs, n- and p-GaP, n- and p-InP, n-ZnO, n-TiO$_2$, and n-SnO$_2$, in aqueous electrolytes is solely determined by the solution pH and shifts in proportion to pH with a slope of -0.059 V/pH. This is explained by the adsorption equilibrium for H$^+$ or OH$^-$ between the semiconductor surface and the solution. For example,

$$S_S - OH + H_{aq}^+ \longleftrightarrow S_S - OH_2^+ \tag{V.1}$$

whereas $S_S - OH$ refers to the $OH$ group present at the semiconductor surface. The $U_{fb}$ for n- and p-Si and that for metal chalcogenides such as n-CdS, n-CdSe and CdTe do not obey the above law, remaining nearly constant in a range of pH range over than about 6 for Si and about 10 for n-CdS.

The synthesis and experiments in this investigation were performed with the intention of developing materials for practical applications. And it can be concluded that each photocatalyst requires specific treatments in order to obtain the best conditions for optimal results in each of the specific applications considering liquid and gas phases.

After the lack of stability observed of different photocatalysts is important to mention the stability as one of the most important paramenters to be considered. In some cases the stability of the new developed materials is forgotten or not analysed in detail. The developed materials sometimes presented in the literature as outstanding, can have different inconvenient side effects after loosing their capacities to decompose other materials. These outstanding materials could be converted into poisonous materials, what must be avoided with the rigorous and proper stability studies.

# VI. Summary

The objective of this investigation was to synthesize diverse semiconductors with intrinsic photocatalytic practical characteristics. Synthesized semiconductors include Ag-$TiO_2$; S-$TiO_2$-$Fe^{3+}$; $In_2Se_3$ and $\beta$-$Ga_2O_3$. These materials were strategically chosen because of their unique energetic applications. Additionally, the aim of this research was to explore an efficient production of high value added materials. Furthermore several experiments in aqueous and/or gas phase were performed to investigate the photocatalytic capability of these materials to be used in the future as photocatalysts focusing on the application of solar energy in order to clean and protect the environment.

Different syntheses of developed photocatalysts were established, and all the properties as well as, characteristics had to be experimentally proven to make sure that the synthesized nanomaterials are stable and photocatalytically active. According to this work, it is possible to confirm, that small modifications in the synthesis can have a big influence in the efficient implementation of the photocatalysts. This means, that even though having obvious and defined information of all the optical properties and a well characterization of the photocatalyst material, predictions about results regarding a visible light active or high efficient photocatalyst cannot be made. It is important to remark that there is no magic photocatalyst, which could be applied in all the cases for the restoration of the environment, as there is no magic medicine to recover the health against all the diseases.

One important parameter considered for the degradation of a model compound with a proper photocatalyst, is the redox potential level, so that a plausible reaction can be thermodynamically established ($\Delta G< 0$).

In the aqueous phase noble elements can be photodeposited on the photocatalyst's surface in order to increase the photonic efficiency. In this case the photodeposited Ag loadings on the $TiO_2$ surface were analyzed. The results show an increase in the photonic efficiency of a commercial $TiO_2$ photocatalyst after consecutively runs, where the photocatalytic activity was recovered by a simple washing technique. Therefore, the reversibility of the poisoning by the chloride ions provides evidence to the assumption, that the recycling of Ag-P25 $TiO_2$ photocatalysts does not have a

# Summary

permanent negative effect on their photocatalytic performance for the degradation of dichloroacetic acid (DCA). The DCA molecule was chosen for these tests because it is a stable compound in aqueous solution which do not absorb light and due to the chlorine ions can be analyzed a possible poisoning effect on the photocatalyst surface by its decomposition can be analyzed. The degradation of DCA is relatively easy to be followed with methods like the total organic carbon (TOC) analysis, chloride electrode measurements and with a pH-stat-titration system by the quantity of released $H^+$. The DCA is also a carcinogenic and toxic compound where can be proved the advantages concerning water remediation treatments with photocatalytic systems. These positive results regarding the recycling of Ag-P25 $TiO_2$ were clearly affected by the preparation procedure minimizing the interaction of the chloride ions with free $Ag^+$ ions or AgOH precipitates in solution and consequently avoiding the formation of AgCl. The formation of AgCl in the system would have lead to complex photolytic reactions, whereby the overall performance for the photodegradation of DCA would have been difficult to predict.

Different pH conditions were also tested when evaluating the photocatalysts photonic efficiency in aqueous phase. The photodeposited silver on $TiO_2$ and the S-doped-$TiO_2$-$Fe^{3+}$ photocatalytic materials showed a higher decomposition of DCA in acidic conditions compared to pH conditions above pH 3. This phenomenon is explained by the high $H^+$ concentration in solution at pH 3 that provides a positively charged surface. However, either by neutral or basic pH the photonic activity showed a decrease in the efficiency. A photocatalytic reaction mainly proceeds near the photocatalyst's surface; therefore interactions between the pollutant and the photocatalyst are governed by the surface charge. Then it is necessary to attach the different model compounds to the photocatalyst by electrostatic charges, so that the compounds are not repulsed but attained to the photocatalysts surfaces.

The increase in conventional energy prices leads to solar energy as an attractive alternative. However photoactive photocatalysts have to be developed for absorbing visible light in order to use it as an alternative energy source. Furthermore the stability of the material should also be defined for real industrial applications in order to avoid any collateral environmental damage. Hence it is important to prove the stability and photocatalytic activity of S-$TiO_2$-$Fe^{3+}$ under visible light irradiation. The mineralization of DCA under visible light takes longer periods of time compared to the UV light

# Summary

experiments. However the capacity of the S-doped $TiO_2$-$Fe^{3+}$ material to absorb visible light with wavelengths higher than 420 nm is of high importance. In correspondence to this attribute, experiments were practiced to test this material also by decomposing substances in the gas phase.

The S-doped $TiO_2$-$Fe^{3+}$ photocatalyst furthermore has the capacity to decompose organic and inorganic compounds in gas phase, as it was successfully determined with $NO_x$ and acetaldehyde respectively under visible light. The same experiments were performed and compared with a commonly standard commercial available photocatalyst $TiO_2$ P25, which is known of its high photoactivity under UV light but not showing without satisfactory results under visible light irradiation.

Consequently and appropriate to the indoor air treatment applications purposes the S-doped $TiO_2$-$Fe^{3+}$ photocatalyst should be considered due to its notable results. Moreover, if this material is stable after several hours of reaction and additionally after consecutively DCA runs, the photocatalyst presented a permanent constant photonic efficiency. In accordance to these results in liquid and gas phases the photocatalyst can be considered as very good being able to take part in the indoor air treatment application as well as for wastewater treatments of recalcitrant and carcinogenic compounds, where the pollutants decomposition results difficult by other methods.

The photonic efficiency calculation for the decomposition of pollutants under visible light is one of the parameters, which should be defined and standardized for comparison between scientific groups and the industry. One relevant point to be considered in this work is the way presented to evaluate the photonic efficiency for visible light according to the light spectrum of the visible lamp.

Regarding the indirect solar energy implementation, the development of suitable materials, which could be efficiently used in solar cells, is required. Considering this, the synthesis of compounds with specific electronic properties for applications related to the solar energy absorption, promotion, operation, utilization and accumulation is necessary.

There are interesting semiconductors with smaller band gaps having good visible light response. However many of these materials decompose easily through photoanodic dissolution of the electrode surface. For example in the case of CdS, the material is oxidized by photogenerated holes so that $Cd^+$ ions dissolve and elemental sulphur deposits on the electrode surface.

By looking into the most efficient materials applied for the solar cell construction $In_2Se_3$ and $Ga_2Se_3$ were the ones fabricated with the highest conversion efficiency, up to 19.2 %. So it was decided to

# Summary

study the synthesis of $In_2Se_3$ semiconductor because of its small band gap, with the aim of increasing its stability by modifying the structure. However, the disadvantage with selenide compounds is the limited stability by heat treatments. In the case of the $In_2Se_3$ photocatalyst the vacancies in the structure could be product of the different syntheses methods. This kind of modification cannot be always previously determined and is almost obtained under unexpected conditions. By characterizing this material with the help of a Kubelka-Munk refraction function values of 3.2 eV and 2.7 eV for the direct and indirect band gap respectively, were calculated. These values are higher than the ones presented in the literature. Furthermore, this material presented a strong crystalline structure and a 2.86 nm particle size average after calcination at 500°C. Even the preliminary adequate light absorption characteristics of the synthesized $In_2Se_3$ powder, was not possible to decompose neither $NO_x$ nor acetaldehyde in gas phase. The values for the redox potential of these two pollutant compounds lay beyond to the $In_2Se_3$ redox potential value. Nevertheless, it was possible to notice an increment of the photocurrent according to the photovoltage measurement. The augment on the photocurrent principally depends on the type of electrolyte used and the amount of semiconductor doped on the electrode to promote a better production of energy. The $In_2Se_3$ material is not stable in aqueous solution as could be observed in the cyclic voltammograms. However, in the literature has been shown, that the mayor success for solar cells systems relies in the combination of $In_2Se_3$ compounds with other material layers, something which is actually being investigated all around the world to increase their light absorption efficiencies.

Due to the latter conditions a further interesting study was the synthesis of a semiconductor with opposite qualities. This means a semiconductor was synthesized having a high stability but a wide gap, and the structure could be modified by heat treatments. Modifications or defects within the semiconductor's structure are suggested as adequate conditions to improve the photonic efficiency. $\beta$-$Ga_2O_3$ appeared to be the adequate material to be studied.

Another relevant and practical use of the solar light is to obtain energy from the water splitting process, where hydrogen can be obtained as a clean energy fuel. In this concern an adequate material is required so that a high efficient production of $H_2$ can be gain from the water splitting. One of the most successfully investigated materials in this field is $Ga_2O_3$. For this reason $\beta$-$Ga_2O_3$ has been synthesized and characterized to observe its photocatalytical properties. The wide band gap of this semiconductor is well known, but after different heat treatments (calcination temperatures of

# Summary

500°C and 1000°C) a defined crystal structure enhanced the photoactivity, which was noticed by the decomposition of acetaldehyde. The importance of the heat treatments in the case of this material has to be mentioned, because just by this experiment it was possible to obtain satisfactory results. When $Ga_2O_3$ powder or films were calcinated at 500°C was not possible to decompose acetaldehyde, but was possible to measure a low photocurrent flow as an electrode configuration.

Still, there are many experiments remaining to be realized with these four different synthesized materials to optimize the use of solar energy. The Ag-P25 $TiO_2$ and the S-doped $TiO_2$-$Fe^{3+}$ photocatalyst could be proposed to prepare coated surfaces and compare the photocatalytic efficiency to make the adoption of these materials into water treatment systems possible.

Because of the lack in time, it was not possible to prepare electrodes with the $\beta$-$Ga_2O_3$ photocatalyst to test the production of hydrogen. However, the problems by synthesizing this photocatalyst, should be pointed out. The first step is to synthesize the gallium acetate, which rate of yield is low. The further reaction to get the appropriate small nanoparticles size of gallium oxide takes one week under constant high temperature conditions, which results in a more complicated synthesis.

In the case of $\beta$-$Ga_2O_3$ none of the found publications could drive to obtain crystalline active nanoparticles able to be used as photocatalyst under UV-A light.

A recommendation would be to make some more experiments with these synthesized materials to find out the level of toxicity. They seem to be stable after photocatalytic reactions, but it should be important to make sure that none of these materials results dangerous for the human beings. In case of human health concern, it is important to know how to treat these hazards.

## VII. References

[1] Braslavsky S. E. and Houk K. N., "Glossary of terms used in photochemistry", *Pure and Applied Chemistry,* Vol. 60, (7), 1055-1106, 1988.

[2] Verhoeven J. W., "Glossary of terms used in photochemistry", *Pure and Application Chemistry,* Vol. 68, (12), 2223-2286, 1996.

[3] Parmon V., Emeline A.V. and Serpone N., "Glossary of terms in photocatalysis and radiocatalysis", *International Journal of Photoenergy,* Vol. 4, 91-131, 2002.

[4] Serpone N. and Salinaro A., "Terminology, relative photonic efficiencies and quantum yields in heterogeneous photocatalysis. Part 1: suggested protoco", *Pure Applied Chemistry,* Vol. 71, 303-320, 1999.

[5] Fujishima A. and Zhang X., "Titanium dioxide Photocatalysis: present situation and future approaches", *Comptes Rendus Chimie,* Vol. 9, 750-760, 2006.

[6] Memming R., "Semiconductor Electrochemistry", WILEY-VCH Verlag GmbH, 2001.

[7] Anpo M., "Use of visible light. Second generation titanium dioxide photocatalysts prepared by the application of an advanced metal ion-implantation method", *Pure and Application Chemistry,* Vol. 72, (9), 1787-1792, 2000.

[8] Takeuchi K., Nakamura I., Matsumoto O., Sugihara S., Ando M. and Ihara T., "Preparation of visible-light-responsive titanium oxide photocatalysts by plasma treatment", *Chemistry Letters,* Vol. 29, (12), 1354-1355, 2000.

[9] Asahi R., Morikawa T., Ohwaki T., Aoki K. and Taga Y., "Visible-Light photocatalysis in nitrogen-doped titanium", *Science,* Vol. 293, 269-271, 2001.

[10] Yu J.C., Yu J., Ho W., Jiang Z. and Zhang L., "Effects of F-doping on the photocatalytic activity and microstructures of nanocrystalline $TiO_2$ powders", *Chemistry of Materials,* Vol. 14, 3808-3816, 2002.

[11] Hong X.T., Wang Z.P., Cai W.M., Lu F., Zhang J., Yang Y.Z., Ma N. and Liu Y.J., "Visible-Light activated nanoparticle photocatalyst of iodine-doped titanium dioxide", *Chemistry of Materials,* Vol. 17, 1548-1552, 2005.

[12] Yu J.C., Zhang L., Zheng Z. and Zhao J., "Synthesis and characterization of phosphated mesoporous titanium dioxide with high photocatalytic activity", *Chemistry of Materials,* Vol. 15, 2280-2286, 2003.

# References

[13] Irie H., Watanabe Y. and Hashimoto K., "Nitrogen-concentration dependence on photocatalytic activity of $TiO_2$-xNx powders", *Journal of Physical Chemistry B,* Vol. 107, 5483-5486, 2003.

[14] Asahi R. and Morikawa T., "Nitrogen complex species ant its chemical nature in $TiO_2$ for visible-light sensitized Photocatalysis", *Chemical Physics,* Vol. 339, 57-63, 2007.

[15] Sakthivel S. and Kisch H., "Daylight photocatalysts by carbon-modified titanium dioxide", *Angewandte Chemie International Edition,* Vol. 42, (40), 4908-4911, 2003.

[16] Ohno T., Mitsui T. and Matsumura M., "Photocatalytic activity of S-doped $TiO_2$ photocatalyst under visible light", *Chemistry Letters,* Vol. 32, 364-365, 2003.

[17] Zhao W., Ma W., Chen C., Zhao J. and Shuai Z., "Efficient degradation of toxic pollutants with $Ni_2O_3/TiO_{2-x}B_x$ under visible irradiation", *Journal of the American Chemical Society,* Vol. 126, (15), 4782-4783, 2004.

[18] Sato S., Nakamura R. and Abe S., "Visible light sensitization of $TiO_2$ photocatalysts by wet method N doping", *Applied Catalysis A,* Vol. 284, 131-137, 2005.

[19] Lettmann C., Hildebrand K., Kisch H., Macyk W. and Maier W.F., "Visible light photodegradation of 4-chlorophenol with a coke-containing titanium dioxide photocatalyst", *Applied Catalysis B: Environmental,* Vol. 32, 215-227, 2001.

[20] Irie H., Watanabe Y. and Hashimoto K., "Carbon-doped anatase $TiO_2$ powders as a visible light sensitive photocatalyst", *Chemistry Letters,* Vol. 32, (8), 772-773, 2003.

[21] Legrini O., Oliveros E. and Braun A.M., "Photochemical Processes for Water Treatment", *Chemical Reviews,* Vol. 93, (2), 671-698, 1993.

[22] Kositzi M., Poulios I., Malato S., Caceres J. and Campos A., "Solar photocatalytic treatment of synthetic municipal wastewater", *Water Research,* Vol. 38, 1147-1154, 2004.

[23] Hoffmann M.R., Scot T.M., Wonyong Ch. and Bahnemann D.W., "Environmental applications of semiconductor photocatalysis", *Chemical Reviews,* Vol. 95, (1), 69-96, 1995.

[24] Bahnemann D.W., "Photocatalytic water treatment: solar energy applications", *Solar Energy,* Vol. 77, 445-459, 2004.

[25] Gerischer H., "Photocatalysis in aqueous solution with small $TiO_2$ particles and the dependence of the quantum yield on particle size and light intensity ", *Electrochimica Acta,* Vol. 40, (10), 1277-1281, 1995.

[26] Kaneko M. and Okura I., "Photocatalysis Science and Technology", Kodansha Springer, 2002.

# References

[27] Kraeutler B. and Bard A. J., "Heterogeneous photocatalytic preparation of supported catalysts Photodeposition of platinum on $TiO_2$ powder and other substrates", *Journal of the American Chemical Society*, Vol. 100, (12), 4317-4318, 1978.

[28] Alivisatos A. P., "Semiconductor clusters, nanocrystals, and quantum dots", *Science*, Vol. 271, (5251), 933-937, 1996.

[29] Brus L. E., "A simple-model for the ionization-potential, electron-affinity, and aqueous redox potentials of small semiconductor crystallites", *Journal of Chemical Physics*, Vol. 79, 5566-5571, 1983.

[30] Brus L. E., "Electron-electron and electron-hole interactions in small semiconductor crystallites - the size dependence of the lowest excited electronic state", *Journal of Chemical Physics*, Vol. 80, 4403-4409, 1984.

[31] Bahnemann D. W., Kormann C. and Hoffmann M. R., "Preparation and characteriza-tion of quantum size zinc oxide: A detailed spectroscopic study", *Journal of Physical Chemistry*, Vol. 91, 3789-3798, 1987.

[32] Wong E. M. and Searson P. C., "ZnO quantum particle thin films fabricated by elec-trophoretic deposition", *Applied Physics Letters*, Vol. 74, 2939-2941, 1999.

[33] Kormann C., Bahnemann D. W. and Hoffmann M. R., "Preparation and characteriza-tion of quantum-size titanium dioxide", *The Journal of Physical Chemistry B*, Vol. 92, 5196-5201, 1988.

[34] Teichner S. J. and Formenti M., "Photoelectrochemistry ", *Photocatalysis and Photoreactors*, Vol. 457-489, 1985.

[35] Friedmann D., Hansing H. and Bahnemann D.W., "Primary Processes During the Photodeposition of Ag Clusters on $TiO_2$ Nanoparticles", *Zeitschrift für Physikalische Chemie*, Vol. 221, 329-348, 2007.

[36] Wang Ch., Böttcher Ch., Bahnemann D.W. and Dohrmann J.K., "In situ electron microscopy investigation of Fe(III)-doped $TiO_2$ nanoparticles in an aqueous environment", *Journal of Nanoparticle Research*, Vol. 6, 119-122, 2004.

[37] Ohno T., Murakami N., Tsubota T. and Nishimura H., "Development of metal cation compound-loaded S-doped $TiO_2$ photocatalysts having a rutile phase under visible light", *Applied Catalysis A*, Vol. 349, 70-75, 2008.

[38] Meissner D., Memming R. and Kastening B., "Photoelectrochemistry of cadmium sulfide. 1. Reanalysis of photocorrosion and Flat-Band Potential", *The Journal of Physical Chemistry*, Vol. 92, (12), 3476-3483, 1988.

[39] Gerischer H. and Mindt W., "The mechanisms of the decomposition of semiconductors by electrochemical oxidation and reduction", *Electrochimica Acta*, Vol. 13, 1329-1341, 1968.

# References

[40] Hidaka H., Honjo H., Horikoshi S. and Serpone N., "Photoinduced $Ag_n^0$ cluster deposition. Photoreduction of $Ag^+$ ions on a $TiO_2$ coated quartz crystal microbalance monitored in real time", *Sensors and Actuators B: Chemical*, Vol. 123, (2), 822-828, 2007.

[41] Tran H., Scott J., Chiang K. and Amal R., "Clarifying the role of silver deposits on titania for the photocatalytic mineralization of organic compounds", *Journal of Photochemistry and Photobiology A: Chemistry*, Vol. 183, 41-52, 2006.

[42] Teoh W.-Y., Mädler L., Beydoun D., Pratsinis S. and Amal R., "Direct (one-step) synthesis of $TiO_2$ and $Pt/TiO_2$ nanoparticles for photocatalytic mineralization of sucrose", *Chemical Engineering Science*, Vol. 60, (21), 5852-5861, 2005.

[43] Zhang F., Guan N., Li Y., Zhang X. and Chen J., "Control of morphology of silver clusters coated on titanium dioxide during photocatalysis", *Langmuir*, Vol. 19, 8230-8234, 2003.

[44] Loginov A.V., Gorbunova V.V. and Boitsova T.B., "Photochemical synthesis and properties of colloidal copper, silver and gold adsorbed on quartz", *Journal of Nanoparticle Research*, Vol. 4, 193-205, 2002.

[45] Ershov B.G., Janata E., Henglein A. and Fojtik A., "Silver atoms and clusters in aqueous solution: absorption spectra and the particle growth in the absence of stabilizing $Ag^+$ ions", *The Journal of Physical Chemistry B.*, Vol. 97, 4589-4594, 1993.

[46] Freund H.J., "Clusters and islands on oxides: from catalysis via electronics and magnetism to optics", *Surface Science*, Vol. 500, (1-3), 271-299, 2002.

[47] Arabatzis I. M., Stergiopoulos T., Bernard M. C., Labou D., Neophytides S. G. and Falaras P., "Silver-modified titanium dioxide thin films for efficient photodegradation of methyl orange", *Applied Catalysis B: Environmental*, Vol. 42, (2), 187-201, 2003.

[48] Xu M.-W., S.-J. Bao and Zhang X-G., "Enhanced photocatalytic activity of magnetic $TiO_2$ photocatalyst by silver deposition", *Materials Letters*, Vol. 59, (17), 2194-2198, 2005.

[49] Zhang Z., "Size-dependent structures and properties of metallic particles and thin films", Ph.D. Dissertation, University of Notre Dame, 2004.

[50] Rao K. M., Pattabi M., Sainkar S. R., Lobo A., Kulkarni S. K., Uchil J. and Murali Sastry M. S., "Preparation and characterization of silver particulate structure deposited on softened poly (4-vinylpyridine) substrates", *Journal Physics. D: Applied Physics*, Vol. 32, 2327-2336, 1999.

[51] Sun H., Zhang Y.Y., Si S.H., Zhua D.R. and Fung Y.S., "Piezoelectric quartz crystal (PQC) with photochemically deposited nano-sized Ag particles for determining cyanide at trace levels in water", *Sensors and Actuators B: Chemical*, Vol. 108, 925-932, 2005.

[52] Bahnemann D.W., Bockelmann D., Goslich R., Hilgendorff M. and Weichgrebe D., in *Trace Metals in the Environment 3: Photocatalytic Purification and Treatment of Water and Air*,

# References

edited by D. F. Ollis H. Al-Ekabi (Hrg.) (Elsevier Science Publishers, Amsterdam), pp. 301-319, 1993.

[53] Wang P., Zakeeruddin S. M., Moser J.-E. and Grätzel M., "A new ionic liquid electrolyte enhances the conversion efficiency of dye-sensitized solar cells", *Journal of Physical Chemistry B*, Vol. 107, 13280-13285, 2003.

[54] Kay A. and Grätzel M., "Dye-sensitized core-shell nanocrystals: improved efficiency of mesoporous tin oxide electrodes coated with a thin layer of an insulating oxide", *Chemistry of Materials*, Vol. 14, (7), 2930-2935, 2002.

[55] Umebayashi T., T. Yamaki, H. Itoh and K. Asai, "Band gap narrowing of titanium dioxide by sulfur doping", *Applied Physics Letters*, Vol. 81, (3), 454-456, 2002.

[56] Ohno T., Akiyoshi M., Umebayashi T., Asai K., Mitsui T. and Matsumura M., "Preparation of S-doped $TiO_2$ photocatalysts and their photocatalytic activities under visible light", *Applied Catalysis A: General*, Vol. 265, 115-121, 2004.

[57] Ohtani B., Ogawa Y. and Nishimoto S., "Photocatalytic activity of amorphous-anatase mixture of titanium (IV) oxide particles suspended in aqueous solutions", *Journal of Physical Chemistry B*, Vol. 101, 3746-3752, 1997.

[58] Zhang Q., Gao L. and Guo J., "Effects of calcination on the photocatalytic properties of nanosized $TiO_2$ powders prepared by $TiCl_4$ hydrolysis", *Applied Catalysis B: Environmental*, Vol. 26, (3), 207-215, 2000.

[59] Gao L. and Zhang Q., "Effects of amorphous contents and particle size on the photocatalytic properties of $TiO_2$ nanoparticles", *Scripta Materialia*, Vol. 44, 1195-1198, 2001.

[60] Arslan I., Akmehmet B. and Tuhkanen T., "Oxidative treatment of simulated dyehouse effluent by UV and near-UV light assisted Fenton's reagents", *Chemosphere*, Vol. 39, 2767-2783, 1999.

[61] Di Paola A., Marci G., Palmisano L., Schiavello M., Uosaki K., Ikeda S. and Ohtani B., "Preparation of polycrystalline $TiO_2$ photocatalyst impregnated with various transition metal ions: characterization and photocatalytic activity for the degradation of 4-nitrophenol", *Journal of Physical Chemistry*, Vol. 106, 637-645, 2002.

[62] Fujishima A., Hashimoto K. and Watanabe T., "$TiO_2$ Photocatalysis Fundamentals and Applications", Bkc. Inc Tokyo Japan, 1999.

[63] Menéndez-Flores V.M., Friedmann D. and Bahnemann D.W., "Durability of Ag-$TiO_2$ photocatalysts assessed for the degradation of dichloroacetic acid", *International Journal of Photoenergy*, Vol. 2008, 11 pages, 2008.

# References

[64] Lindner M., Bahnemann D.W., Hirthe B. and Griebler W.-D., "Solar Water Detoxification: Novel $TiO_2$ Powders as Highly Active Photocatalysts", *Transactions of the ASME*, Vol. 119, 120-125, 1997.

[65] Negishi N., Takeuchi K. and Ibusuki T., "The surface structure of titanium dioxide thin film photocatalyst", *Applied Surface Science*, Vol. 121-122, 417-420, 1997.

[66] Ohko Y., Tryk D.A., Hashimoto K. and Fujishima A., "Autoxidation of acetaldehyde initiated by $TiO_2$ photocatalysis under weak UV illumination", *Journal of Physical Chemistry B*, Vol. 102, (15), 2699-2704, 1998.

[67] Lide D.R., *Handbook of Chemistry and Physics*, 70th ed. (Ed. CRC, Boca Raton, FL. USA), 1990.

[68] Bahnemann D.W., Bockelmann D. and Goslich R., "Mechanistic studies of water detoxification in illuminated $TiO_2$ suspension", *Solar Energy Materials*, Vol. 24, (1-4), 564-583, 1991.

[69] Southampton Electrochemistry Group, "Instrumental methods in electrochemistry", Ellis Horwood, 1985.

[70] Marugán J., Hufschmidt D., Sagawe G., Selzer V. and Bahnemann D.W., "Optical density and photonic efficiency of silica-supported $TiO_2$ photocatalysts", *Water Research*, Vol. 40, 833-839, 2006.

[71] Bahnemann D.W., "Ultrasmall metal oxide particles: preparation, photophysical characterization and photocatalytic properties", *Israel Journal of Chemistry*, Vol. 33, 115-136, 1993.

[72] Vamathevan V., "The role of silver deposits on titania in the photocatalytic oxidation of organics in aqueous media", PhD Dissertation, The University of New South Wales, 2003.

[73] Chan S. C. and Barteau M. A., "Preparation of highly uniform $Ag/TiO_2$ and $Au/TiO_2$ supported nanoparticle catalyst by photodeposition", *Langmuir*, Vol. 21, 5588-5595, 2005.

[74] Robert D. and Malato S., "Solar photocatalysis: a clean process for water detoxification", *The Science of the Total Environment*, Vol. 291, 85-97, 2002.

[75] Muñoz I., Peral J., Ayllon J., Malato S., Passarinho P. and Domenech X., "Life cycle assessment of a couple solar photocatalytic-biological process for wastewater treatment", *Water Research*, Vol. 40, 3533-3540, 2006.

[76] Ohno T., Miyamoto Z., Nishijima K., Kanemitsu H. and Xueyuan F., "Sensitization of photocatalytic activity of S- or N-doped $TiO_2$ particles by adsorbing $Fe^{3+}$ cations", *Applied Catalysis A: General*, Vol. 302, 62-68, 2006.

# References

[77] Tauc J., Grigorovici R. and Vancu A., "Optical properties and electronic structures of amorphous germanium", *Physica Status Solidi*, Vol. 15, 627-637, 1966.

[78] Li Q., Xie R., Shang J.K. and Mintz E.A., "Effect of precursor ratio on synthesis and optical absorption of TiON photocatalytic Nanoparticles", *Journal of the American Ceramic Society*, Vol. 90, (4), 1045-1050, 2007.

[79] Rupa A.V., Divakar D. and Sivakumar T., "Titania and noble metals deposited titania catalysis in the photodegradation of tartazine", *Cataysis Letters*, Vol. 132, 259-267, 2009.

[80] Znad H. and Kawase Y., "Synthesis and characterization of S-doped Degussa P25 with application in decolorization of orange II dye as a model substrate", *Journal of Molecular Catalysis A: Chemical*, Vol. 314, 55-62, 2009.

[81] Enright B. and Fitzmaurice D., "Spectroscopic determination of electron and hole effective masses in a nanocrystalline semiconductor film", *Journal of Physical Chemistry B*, Vol. 100, (3), 1027-1035, 1996.

[82] Sterner J., Malmstrom J. and Stolt L., "Study on ALD $In_2S_3$/Cu(In,Ga)$Se_2$ interface formation", *Progress in Photovoltaics*, Vol. 13, (3), 179-193, 2005.

[83] Naghavi N., Spiering S., Powalla M., Cavana B. and Lincot D., "High-Efficiency Copper Indium Gallium Diselenide (CIGS) Solar Cells with Indium Sulfide Buffer Layers Deposited by Atomic Layer Chemical Vapor Deposition (ALCVD)", *Progress in Photovoltaics*, Vol. 11, (7), 437-443, 2003.

[84] Avivi S., Palchik O., Palchik V., Slifkin M. A., Weiss A. M. and Gedanken A., "Sonochemical Synthesis of Nanophase Indium Sulfide", *Chemistry of Materials*, Vol. 13, (6), 2195-2200, 2001.

[85] Afzaal M., Crouch D., O´Brien P. and Park J. H., "Metal-organic chemical vapor deposition of beta-$In_2S_3$ thin films using a single-source approach", *Journal of Materials Science: Materials in Electronics*, Vol. 14, (9), 555-557, 2003.

[86] Nagesha D. K., Liang X. R., Mamedov A. A., Gainer G., Eastman M. A., Giersig M., Song J. J., Ni T. and Kotov N. A., "$In_2S_3$ nanocolloids with excitonic emission: $In_2S_3$ vs CdS comparative study of optical and structural characteristics", *The Journal of Physical Chemistry B*, Vol. 105, (31), 7490-7498, 2001.

[87] Revaprasadu N., Malik M. A., J. Carstens and P. O´Brien, "Novel single-molecule precursor routes for the direct synthesis of InS and InSe quantum dots", *Journal of Materials Chemistry*, Vol. 9, 2885-2888, 1999.

[88] Bindu K., Kartha C. S., Vijayakumar K. P., T. Abe and Kashiwaba Y., "Structural optical and electrical properties of $In_2Se_3$ thin films formed by annealing chemically deposited Se and vacuum evaporated In stack layers", *Applied Surface Science*, Vol. 191, 138-147, 2002.

# References

[89] Lu K., Sui M. L., Perepezko J. H. and Lanning B., "The kinetics of indium/amorphous-selenium multilayer thin film reactions", *Journal of Materials Research*, Vol. 3, 771-779, 1999.

[90] Afzaal M., Crouch D. and O´Brien P., "Metal-organic chemical vapor deposition of indium selenide films using a single-source precursor", *Materials Science and Engineering B*, Vol. 116, (3), 391-394, 2005.

[91] Choi I. H. and Yu P. Y., "Properties of phase-pure InSe films prepared by metalorganic chemical vapor deposition with a single-source precursor", *Journal of Applied Physics*, Vol. 93, (8), 4673-4677, 2003.

[92] Rabchynski S. M., Ivanou D. K. and Streltsov E. A., "Photoelectrochemical formation of indium and cadmium selenide nanoparticles through Se electrode precursor", *Electrochemistry Communications*, Vol. 6, (10), 1051-1056, 2004.

[93] Uosaki K., Kaneko S., Kita H. and Chevy A., "Electrochemical behavior of p-type indium selenide single crystal electrodes in dark and under illumination", *The Chemical Society of Japan*, Vol. 59, 599-605, 1986.

[94] Singh K., Saxena N. S., Srivastav O. N., Patidar D. and Sharm T. P., "Energy band gap of $Se_{100-x}In_x$ chalcogenide glasses", *Chalcogenide Letters*, Vol. 3, (3), 33-36, 2006.

[95] Pathan H.M., Kulkarni S.S., Mane R.S. and Lokhande C.D., "Preparation and characterization of indium selenide thin films from a chemical route", *Materials Chemistry and Physics*, Vol. 93, 16-20, 2005.

[96] Emziane M. and Le Ny R., "Crystallization of $In_2Se_3$ semiconductor thin films by post-deposition heat treatment. Thickness and substrate effects", *Journal of Physics D: Applied Physics*, Vol. 32, 1319-1328, 1999.

[97] Malik M. A., O'brien P. and Revaprasadu N., "A novel route for the preparation of CuSe and $CuInSe_2$ nanoparticles", *Advanced Materials*, Vol. 11, (17), 1441-1444, 1999.

[98] Yang S. and Kelley D. F., "The Spectroscopy of InSe Nanoparticles", *The Journal of Physical Chemistry B*, Vol. 109, 12701-12709, 2005.

[99] Tabernor J., Christian P. and O´Brien P., "A general route to nanodimensional powders of indium chalcogenides", *Journal of Materials Chemistry*, Vol. 16, 2082-2087, 2006.

[100] Hogg J.H.C., "The crystal structure of $In_6Se_7$", *Acta Crystallographica Section B*, Vol. 27, 1630-1634, 1971.

[101] Memming R., "In: B. Conway, J.O'M Boekris and E. Yeager, Editors, Comprehensive Treatise of Electrochemistry", Vol. 7, 529, 1983.

# References

[102] Baumanis C. and Bahnemann D.W., "$TiO_2$ Thin film electrodes: correlation between photocatalytic activity and electrochemical properties", *The Journal of Physical Chemistry C*, Vol. 112, (48), 19097-19101, 2008.

[103] Memming R., "Energy production by photoelectrochemical processes", *Philips Technical Review*, Vol. 38, (6), 160-177, 1978-1979.

[104] Chandra S., Singh D. P., Srivastava P. C. and Sahu S. N., "Electrodeposited semiconducting molybdenum selenide films: II. Optical, electrical, electrochemical and photoelectrochemical solar cell studies", *Journal of Physics D: Applied Physics*, Vol. 17, 2125-2138, 1984.

[105] Geller S., "Crystal structure of $\beta$-$Ga_2O_3$", *The Journal of Chemical Physics*, Vol. 33, (3), 676-684, 1960.

[106] Foster L. M. and Stumpf H. C., "Analogies in the gallia and alumina systems. The preparation and properties of some low-alkali gallates", *Journal of the American Chemical Society*, Vol. 73, (4), 1590-1595, 1951.

[107] Zhao J. G., Zhang Z. X., Ma Z. W., Duan H. G., Guo X. S. and Xie E. Q., "Structural and photoluminescence properties of $\beta$-$Ga_2O_3$ nanofibres fabricated by electrospinning method", *Chinese Physics Letters*, Vol. 25, (10), 3787-3789, 2008.

[108] Ogita M., Saika N., Nakanishi Y. and Hatanaka Y., "$Ga_2O_3$ thin films for high-temperature gas sensors", *Applied Surface Science*, Vol. 142, 188-191, 1999.

[109] Zhao B. and Zhang P., "Photocatalytic decomposition of perfluorooctanoic acid with $\beta$-$Ga_2O_3$ wide bandgap photocatalyst", *Catalysis Communications*, Vol. 10, 1184-1187, 2009.

[110] Sakata Y., Matsuda Y., Yanagida T., Hirata K., Imamura H. and Teramura K., "Effect of metal ion addition in a Ni supported $Ga_2O_3$ photocatalyst on the photocatalytic overall splitting of $H_2O$", *Catalysis Letters*, Vol. 125, 22-26, 2008.

[111] Funk H. and Paul A., "Über Umsetzungen des Galliumchlorids mit organischen Verbindungen", *Zeitschrift für anorganische und allgemeine Chemie*, Vol. 330, 70-77, 1964.

[112] Sinha G., Adhikary K. and Chaudhuri S., "Optical properties of nanocrystalline $\alpha$-GaO(OH) thin films", *Institute of Physics: Condensed Matter*, Vol. 18, 2409-2415, 2006.

[113] Jiang H., Chen Y., Zhou Q., Su Y., Xiao H. and Zhu L., "Temperature dependence of $Ga_2O_3$ micro/nanostructures via vapor phase growth", *Materials Chemistry and Physics*, Vol. 103, 14-18, 2007.

[114] Khan A., Jadwisienczak W.M. and Kordesch M.E., "One-step preparation of ultra-wide $\beta$-$Ga_2O_3$ microbelts and their photoluminescence study", *Physica E*, Vol. 35, 207-211, 2006.

# References

[115] Hajnal Z., Miró J., Kiss G., Réti F., Deák P., Herndon R. C. and Kuperberg J. M., "Role of oxygen vacancy defect states in the n-type conduction of $\beta$-Ga$_2$O$_3$", *Journal of Applied Physics*, Vol. 86, (7), 3792-3796, 1999.

[116] Binet L. and Gourier D., "Origin of the blue luminiscence of $\beta$-Ga$_2$O$_3$", *Journal of Physics and Chemistry of Solids*, Vol. 59, (8), 1241-1249, 1998.

[117] Liang C. H., Meng G. W., Wang G. Z., Wang Y. W. and Zhang L. D., "Catalytic synthesis and photoluminescence of $\beta$-Ga$_2$O$_3$ nanowires", *Appied Physics Letters*, Vol. 78, (21), 3202-3204, 2001.

[118] Landolt-Börnstein, "Semiconductors: Physics of Non-Tetrahedrally Bonded Elements & Binary Compounds II", Vol. III, (17f), 293, 1982.

[119] Fu D. and Kang T. W., "Electrical properties of gallium oxide grown by photoelectrochemical oxidation of GaN epilayers", *The Japan Society of Applied Physics*, Vol. 41, 1437-1439, 2002.

[120] Passlack M., Schubert E. F., Hobson W. S., Hong N., Moriya M., Chu S. N. G., Konstadinidis K., Mannaerts J. P., Schnoes M. L. and Zydzik G. J., "Ga$_2$O$_3$ films for electronic and optoelectronic applications", *Journal of Applied Physics*, Vol. 77, (2), 686-693, 1994.

# VIII. Annex 1

VIII.1. Halogen lamp spectrum for visible light emission

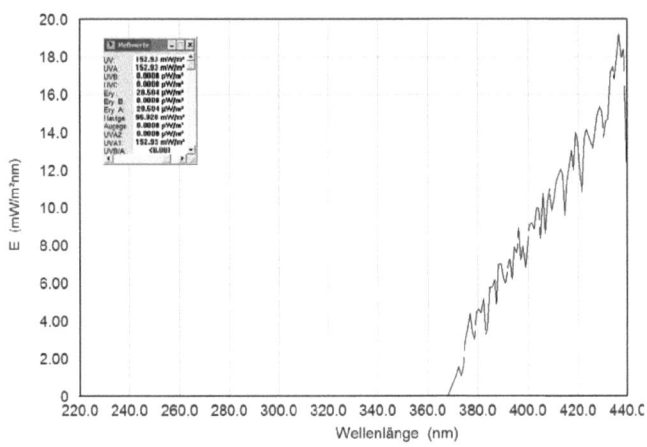

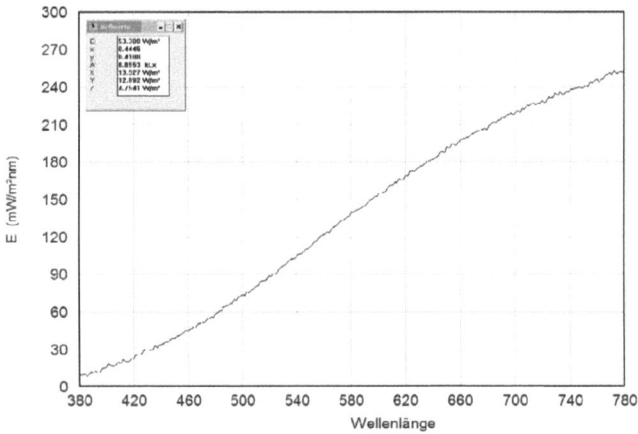

Die VDM Verlagsservicegesellschaft sucht für wissenschaftliche Verlage abgeschlossene und herausragende

## Dissertationen, Habilitationen, Diplomarbeiten, Master Theses, Magisterarbeiten usw.

für die kostenlose Publikation als Fachbuch.

Sie verfügen über eine Arbeit, die hohen inhaltlichen und formalen Ansprüchen genügt, und haben Interesse an einer honorarvergüteten Publikation?

Dann senden Sie bitte erste Informationen über sich und Ihre Arbeit per Email an *info@vdm-vsg.de*.

**Sie erhalten kurzfristig unser Feedback!**

VDM Verlagsservicegesellschaft mbH
Dudweiler Landstr. 99　　　　　　　Telefon　+49 681 3720 174
D - 66123 Saarbrücken　　　　　　　Fax　　　 +49 681 3720 1749
**www.vdm-vsg.de**

Die VDM Verlagsservicegesellschaft mbH vertritt

Printed by Books on Demand GmbH, Norderstedt / Germany